The Voice of Nature in Ted Hughes's Writing for Children

Despite the fame Ted Hughes's poetry has achieved, there has been surprisingly little critical writing on his children's literature.

This book identifies the importance of Hughes's children's writing from an ecocritical perspective and argues that the healing function that Hughes ascribes to nature in his children's literature is closely linked to the development of his own sense of environmental responsibility. This book will be the first sustained examination of Hughes's greening in relation to his writing for children, providing a detailed reading of Hughes's children's literature through his poetry, prose and drama as well as his critical essays and letters. In addition, it also explores how Hughes's children's writing is a window to the poet's own emotional struggles, as well as his environmental consciousness and concern to reconnect a society that has become alienated from nature.

This book will be of great interest to not only those studying Ted Hughes, but also students and scholars of environment and literature, ecocriticism, children's literature and twentieth-century literature.

Lorraine Kerslake holds a PhD in children's literature and ecocriticism, and has published widely in the field. She also has a BA in English and French studies and an MA in Translation and Interpreting from the University of Alicante, Spain, where she currently works teaching English Language and Literature.

Routledge Environmental Humanities
Series editors: Iain McCalman and Libby Robin

The *Routledge Environmental Humanities* series is an original and inspiring venture recognising that today's world agricultural and water crises, ocean pollution and resource depletion, global warming from greenhouse gases, urban sprawl, overpopulation, food insecurity and environmental justice are all *crises of culture*.

The reality of understanding and finding adaptive solutions to our present and future environmental challenges has shifted the epicenter of environmental studies away from an exclusively scientific and technological framework to one that depends on the human-focused disciplines and ideas of the humanities and allied social sciences.

We thus welcome book proposals from all humanities and social sciences disciplines for an inclusive and interdisciplinary series. We favour manuscripts aimed at an international readership and written in a lively and accessible style. The readership comprises scholars and students from the humanities and social sciences and thoughtful readers concerned about the human dimensions of environmental change.

The Voice of Nature in Ted Hughes's Writing for Children

Correcting Culture's Error

Lorraine Kerslake

Routledge
Taylor & Francis Group
LONDON AND NEW YORK

earthscan
from Routledge

First published 2018
by Routledge

2 Park Square, Milton Park, Abingdon, Oxfordshire OX14 4RN
52 Vanderbilt Avenue, New York, NY 10017

Routledge is an imprint of the Taylor & Francis Group, an informa business

First issued in paperback 2019

British Library Cataloguing-in-Publication Data
A catalogue record for this book is available from the British Library

Library of Congress Cataloging-in-Publication Data
A catalog record has been requested for this book

ISBN: 978-1-138-57367-3 (hbk)
ISBN: 978-0-367-85600-7 (pbk)

Typeset in Goudy
by Wearset Ltd, Boldon, Tyne and Wear

Contents

Acknowledgements

I am greatly indebted to many people for their help and guidance during the writing of this book. First, to Terry Gifford for his generosity, encouragement and continuous support. My gratitude is extended to many scholars, especially Lissa Paul, Mick Gowar, Claas Kazzer and Ann Skea, whose kindness and generous observations were of great help. My thanks also to the curators, librarians and staff at the British Library and Emory University, Atlanta. I am also grateful to the Ted Hughes Society for giving me the opportunity to first publish the early seeds of my research and having had the chance to meet and share time with such a warm and generous group of scholars during the conferences hosted over the past years. Last, but never least, to my husband for his patience and love along this journey.

Introduction

Despite the fame Ted Hughes's poetry has achieved, there has been surprisingly little critical writing on his children's literature. Given that Hughes thought that children were more open and receptive than adults, both to the subconscious as well as to nature, it was through his children's work that he explored a part of his inner self that was often masked in his writing for adults. Hughes started writing for children early in his literary career, beginning in Benidorm on the honeymoon he spent with Sylvia Plath in 1956 and published over 25 books for children in the form of poetry, prose, theatre and critical essays. Following the early critical work of Keith Cushman (1983), both Claas Kazzer (1999) and Lissa Paul (1986, 2000), followed by Neil Roberts (2006), Michael Lockwood (2009) and more recently Mick Gowar (2011) and David Whitley (2011) have offered critical readings of Hughes's children's works. However, there is at present no full-length study which examines in detail Hughes's contribution to the field of children's literature.

This book takes a fresh look at Hughes's work by identifying the importance of his writing for children, from an ecocritical perspective, as part of a major professional project which Hughes embraced throughout his entire career. *The Voice of Nature in Ted Hughes's Writing for Children* argues that the healing function that Hughes ascribes to nature in his children's literature is closely linked to the development of his own sense of environmental responsibility and explores how Hughes articulates human relationship with nature and the sense of 'ecological healing' that is portrayed in the realm of his children's writing.

Hughes's healing quest reveals itself as being two-fold. On the one hand it was through his writing for children that Hughes explored a part of his inner self that was often less explicit in his writing for adults, embodying a hidden autobiographical self closely related to Hughes's own emotional struggle and personal crisis in the years following the traumatic suicides of Sylvia Plath and Asia Wevill, together with Shura. However, in a more general sense, his healing quest also reads as voicing his own environmental consciousness and concern to reconnect a society that has become more alienated from nature, resulting in our current global environmental crisis.

In order to frame the concept of healing in Hughes's children's writing, this study has been structured in two parts. The literary analysis of Hughes's children's writing that shapes Part II is nourished by the background discussion in Part I, that focuses in Chapter 1 on the necessary biographical aspects of Hughes's childhood and early engagement with the natural world, and is followed in Chapter 2 by his greening and development as an environmental writer ahead of his time. This has been contextualised in Chapter 3 by looking at his own critical essays, as well as his handbook on writing poetry, *Poetry in the Making* (1967), together with his collected letters and other unpublished material from the archives at the British Library, in order to gain an insight into the theoretical and critical background that make up the cauldron of Hughes's own ideas related to imagination, writing, art and education – concepts which are all central to Hughes's concept of healing. Chapter 4 analyses Hughes's work as an eco-playwright and looks at the ecological implications that appear in his *Collected Plays for Children* (2001). Following this, Hughes's version of 'Orpheus' (1971) reads as the most redemptive of his children's plays and that which best illustrates the achievement of his personal shamanic quest of healing. Given the daunting weight and scope of Hughes's poetic oeuvre, Chapter 5 has surveyed the development of Hughes as an eco-poet, offering a chronological analysis of his work and focusing on those poems which best illustrate Hughes's healing quest as well as exploring Hughes's artistic collaboration with Leonard Baskin and Reg Lloyd. Chapter 6 looks at Hughes as an eco-storyteller and analyses *Ffangs the Vampire Bat and the Kiss of Truth* (1986) as a redemptive story. It also looks at the range of Hughes's 'Creation Tales' and closes by reading *The Iron Man* (1968) and *The Iron Woman* (1993) as healing myths, which are largely informed by Hughes's own social and political interventions related to environmental concerns such as river pollution and the dumping of toxic waste.

Reconnecting with nature

Hughes's children's writing draws on the need for humans, especially children, to reconnect with nature. Throughout his life Hughes aspired to teach children to respect the natural world through a positive connection to nature.

Similar concerns have been recently voiced about the way today's indoor generation of high-tech children are becoming increasingly divorced from nature.[1] In 2005, Richard Louv coined the term 'nature-deficit disorder' to refer to today's generation of children being deprived from nature and warned about the consequences this would have on the future stewardship of our environment. However, ever since the mid-twentieth century other voices have argued that a child's direct experience of nature exerts a crucial effect on their development. Both Edith Cobb ('The Ecology of Imagination in Childhood', 1959), and Rachel Carson ('Help Your Child to Wonder, 1965 [1956])[2] understood the importance of bonding with nature in early childhood. Cobb's work proposed that contact with nature stimulates creativity and imagination and put forth similar views to those of ecopsychology.

In the 1980s E. O. Wilson (1984) coined the term biophilia, a 'love of nature' that can be linked to the emotional connection we form with nature as children, followed by Stephen Kellert (1993) whose biophilia hypothesis suggests that humans are innately affiliated with nature. Sadly, in Western society today, many children have few opportunities for any contact with the natural world. As Kellert has said, our culture has become 'so estranged from its natural origins it has failed to recognize our species' basic dependence on nature as a condition of growth and development' (1993: 118).

As Hughes was aware, this disconnection from the natural world was just as great in adults. He was fully conscious of the existential threats from our environmental crisis in the form of climate change and pollution, the loss of natural habitat and biodiversity as well as the state of the oceans and freshwater resources. In response to the awareness of the environmental crisis which began in the last century, Hughes's work and his own environmental consciousness explore the rifts that have taken place between humankind and the natural world in order to foster a sense of interconnectedness between humans and nature. Hughes's work advocates the need for environmental conservation which can be traced throughout his literary career. The poet's sense of nature, together with his own environmental consciousness, were already latent in his early childhood experiences close to the natural world surrounding him. Hughes's environmental awareness, together with his strong sense of ecological responsibility, can be seen throughout his writing and work, beginning with his early animal poems which show a preoccupation for the conservation of local species, and throughout his children's works where his love of the natural world and his desire to reconnect with nature are often expressed through his ongoing concern for environmental education.

Ecocriticism and environmental awareness

Hughes's own growing environmentalism, together with the convictions that underlie his poetic voice and the role he took on to communicate and heal human alienation from nature, call for an ecocritical reading of his writing. Given that ecocriticism engages with the intellectual and ethical challenges of environmental issues raised by a reading of literature, in the case of Hughes's children's literature it offers a theoretical approach to reading Hughes that gives insight into the way in which he articulates human alienation from nature and hopes to resolve it in his children's writing.

For ecocriticism, questioning the ways in which children's texts contribute to attitudes towards nature and environment and looking at how the child is represented textually as well as the way they are connected with nature is a critical aspect of inquiry (Dobrin and Kidd, 2004). Hughes's vision extends the representation of nature in children's literature to a greater degree by offering the child guidelines, values and expectations of the adult world. At the same time, his writing raises important questions for ecocriticism, such as: How does his children's writing relate to our ecological crisis and what can an ecocritical reading of his work reveal? What is the evidence for its application? Is Hughes's children's language sensitive to environmental issues? Does it contribute to a shift in sensibility? Does it raise a response in readers? What questions concerning environmental issues does it raise? According to Hughes, what is needed to re-establish the balance and harmony between nature and humankind? In which works is this balance achieved and in which works does it fail? From an aesthetic point of view how *good* is his children's writing? And, underlying the central question of this book, is there a healing quest successfully embedded in his children's writing?

Despite the fact that most major critics, including Neil Roberts (*Ted Hughes: A Literary Life*, 2006) and Keith Sagar (*Literature and the Crime Against Nature*, 2005), have now recognised that Hughes's poetic work articulates and reflects his environmental concerns, very few critics have looked at the development of Hughes's greening which can be traced through his writing and educational work for children, and which took place parallel to the cultural development of environmental activism and awareness that can be dated in Britain to the second half of the post-war period. One of the first critics to note that 'Hughes's goddess has an ecological aspect, as well as a psychological and religious [one]' was Craig Robinson in *Ted Hughes as Shepherd of Being* (1989: 208). However, the first critics to make a proper case for Hughes as an ecological poet and offer an ecocritical reading of Hughes work were Leonard M. Scigaj (1991) and Terry Gifford (2011 [1995]). Whilst Craig Robinson's book is rather simplistic and focuses on Hughes's first works, Leonard M. Scigaj was the first to analyse Hughes's sense of ecological responsibility in *Ted Hughes* (1991), and ecocritic Terry Gifford in *Green Voices* (2011 [1995]) was the first to compare and analyse the notions and constructions of nature that underpin Hughes's work. Over the last decade other major critics have recognised Hughes's engagement with our environmental crisis, including the pioneer of Hughes scholarship, Keith Sagar as well as Neil Roberts – who talks of a religious and ultimately ecological responsibility underlying Hughes's *River* poems (see Roberts 2006). More recently Yvonne Reddick in *Ted Hughes: Environmentalist and Ecopoet* (2017) has shed light on how Hughes's ecopoetry reads as reflecting his environmental awareness.

Hughes's shamanic quest

Hughes writes for a world that has lost its balance. According to the poet, 'the story of the mind exiled from Nature is the story of Western Man' (*WP*: 129). Many authors (see Campbell, 1983; Eliade, 1964; Melson, 2001; Shepard, 1996) have argued that historically humans were once both psychologically and physically closer to nature than they are now. Consequently, one of the greatest sources of culture clash between European colonists and indigenous cultures has been their different attitudes towards nature. The roots of modern civilisation's estrangement from the natural world lie in the same malaise registered ideologically by the Romantics, indicative of a deeper shift in the socioeconomic sphere, and which Marx understood as being symptomatic of the rise of capitalism. For many non-Western cultures human alienation from nature has resulted in a sickness of the soul.

Hughes records this epic journey in his writing and explores the spiritual technique known as shamanism as a healing force to emphasise the holiness of nature and restore the equilibrium. Hughes believed that science and technology had repressed human's imagination and emotional needs and, as a consequence, society had lost its imagination and ability to empathise with others. Following this, his creative universe can be read as a context for healing the human/nature split, and an attempt to cure Western society's alienation from nature by unlocking our senses and reconnecting culture to its roots. Hughes regards shamanism as a force of reintegration and places great emphasis on the poetic experience to restore the balance. It is in this context that the poet develops his shamanic quest, where the shaman/poet acts as a storyteller, healer and mediator speaking in a language that is in tune with the natural world. For Hughes, when literature is working at its best it has a renovating and therapeutic effect, creating a healing process and mediating between the reader and the outside world. So the true function of poetry (and here we can include his prose as an extension of his verse) is not just to heal the fractured self of the poet/artist, but in a wider and more social sphere Hughes believed that poetry could work on others too. The poet acts as a literary shepherd in the sense that his belief in the affective power of words is grounded in his goal of attaining an ecological balance. Given that Hughes consistently draws on Robert Graves's *The White Goddess* (1948), his sense of nature is also closely linked to his evocation of the goddess of nature.

The healing function of nature is closely connected to other concepts that Hughes explores throughout his children's work: myth, imagination and education. These themes are closely embedded in his approach to his children's writing and form part of the universe of Hughes's symbolic framework that are best expressed in the realm of children's literature. In addition, the healing process that underlies Hughes's writing is best achieved through the creative development of his children's work and the 'upbeatness' of his children's literature which exercises that healing function as being redemptive. Furthermore, for Hughes, children constitute an ideal audience since they are not yet conditioned by society. Given its simplicity and audience, children's literature is especially effective in allowing the reader to penetrate magical worlds full of symbols, images and archetypes and cross the borders between the worlds of dreams and fantasy.

The representation of nature

In the case of Hughes, the concern for nature that emerges so distinctly in his children's literature is part of a major professional project which the poet embraced throughout his entire career. Given its importance, any study of Hughes must engage with how the poet develops his concept of nature and portrays the engagement and interconnectedness between human nature and external nature.

Hughes redefines the idea of 'nature', by looking at its different representations in the nature/culture dichotomy, at times both excluding and including the human, and by questioning the interactions that take place between the human and the non-human. For nature is, more than a key word, as much a concept as a reality that we are both part of yet, at the same time, apparently where we feel we cannot be wholly at home in a sustainable unproblematic relationship. Not only does Hughes *challenge* the complex notion of nature through the way he observes, conceives and depicts it, but his actual sense of nature progresses throughout his work, drawing on Rousseau's sense of nature, as an innate disposition, and one which we are born with and develop throughout our life in response to our childhood environment. As Paul Bentley (1998) has observed, Hughes's early 'social and political' sense of nature as portrayed in his adult works *The Hawk in the Rain* (1957) and *Lupercal* (1960) is different from that portrayed in *Moortown Diary* (1989) and *Remains of Elmet* (1979). In line with Bentley, *Moortown Diary* and Hughes's own hands-on experience as a farmer reads as being the turning point of his sense of ecological responsibility and lyrical duty as a poet. Moreover, as Gifford has correctly stated,

> By the time of *Remains of Elmet* (1979) Hughes had come to regard human culture as nature – industry, farms and people not only embedded in nature in the Calder Valley of West Yorkshire, but suffering the same processes of growth and decay
>
> (2011 [1995]: 81)

so that the metaphor of nature is represented as culture, whilst culture is 'so deeply embedded in nature that nature can also be represented as culture' (Gifford, 2011 [1995]: 81). So both the experience of nature as well as the conception of nature itself is always mediated by culture. As French philosopher and sociologist Edgar Morin has put it, 'All that speaks of nature speaks of society. The "conquest of nature", the "return to nature" are the most social of social ideas' (1982: 130).

So the concept of 'nature' in Hughes's poetic universe is complex in relation to the philosophical human–nature relationship. Kate Soper in *What is Nature?* (1995) questions this notion as a cultural construction based on anthropocentrism. Soper's definition of nature as a social construction and 'the concept through which humanity thinks its difference and specificity' (1995: 155) provides a useful starting point. Soper implicitly draws on the ecofeminist critique of a rationalist dualism (see Plumwood, 1993) in which the relationship between the human and the non-human is still the most fundamental of questions – so fundamental that Western thinking completely depends on it. Terry Gifford's concept of 'post-pastoral'[3] (where 'post-' does not mean 'after' but implies instead 'reaching beyond' the constraints of idealised pastoral poetry) is also particularly valuable as a working framework to analyse Hughes's nature poetry (1994, 2011 [1995]).[4]

As will be revealed, throughout his career, and as part of his overall project, Hughes's sense of nature becomes more holistic. Instead of writing about a physical nature that is untouched and separate from a human being, his poetry bridges the gap between the human and the non-human allowing the reader to cross the boundaries between both worlds. Using Timothy Clark's definition of ecopoetry, Hughes's poems are 'conceived as a space of subjective redefinition and rediscovery through encounters with the non-human' (2011: 139). Hughes reads as an eco-poet rather than a nature poet not because his poetry is just poetry adorned with flowers and insects or animals or trees, but because of his life-long preoccupation with environmental issues as real subjects that both underlie and affect human ecology embedded in nature. It is precisely because of his perception of what David Abram (1996) and most ecocritics now call the 'more than human',[5] and the way in which he goes beyond the realm of the human in his attempt to describe nature from a non-anthropocentric perspective, that his poetry helps to create the recognition of our interdependency on the biosphere.

In 1996, in an interview with Eliat Negev,[6] Hughes said that: 'Every work of art stems from a wound in the soul of the artist. When a person is hurt, his immune system comes into operation and the self-healing process takes place, mental and physical' (quoted in Sagar, 2006: xi). In the case of Hughes, the healing energy of his work tried to alleviate 'the explosion of pain' (Sagar, 2006: xi) following the tragedies that occurred in his life. As this book reveals, it would be that same voice of pain that Hughes strove to reconcile through his ongoing shamanic quest for healing, and which he would most successfully resolve in the realm of his writing for children.

Notes

1 In 2012, a National Trust report called *Natural Childhood* (Moss, 2012) revealed the growing gap between children and nature. In 2013, the Royal Society for the Protection of Birds published a study which concluded that four out of five children in the UK were not adequately 'connected to nature' (RSPB, 2013).

2 Carson's essay 'Help Your Child to Wonder' first appeared in *Woman's Home Companion* in July 1956 and was later published posthumously by Harper & Row as *The Sense of Wonder*, in 1965.

3 Gifford's notion of 'post-pastoral' looks at how a writer responds to and challenges pastoral conventions (Gifford, 2011 [1995]), through six key questions that a work might raise for the reader:

   1 Can awe in the face of nature (e.g. landscapes) lead to humility in our species?
   2 What are the implications of recognising that we are part of nature's creative-destructive processes?
   3 If our inner nature echoes outer nature, how can the outer help us understand the former?
   4 If nature is culture, is culture nature?
   5 How can our distinctively human consciousness, which gives us a conscience, be used as a tool to heal our alienation from nature?
   6 How should we address the ecofeminist insight that the exploitation of our planet is aligned with our exploitation of the less powerful?

4 See Gifford's chapter, 'Laureate of Nature: The Poetry of Ted Hughes' in *Green Voices* (2011: pp. 132–156); Gifford (1994: pp. 129–141).

5 The phrase 'More than human' was first used by Abram in *The Spell of the Sensuous* (1996) to overcome the nature/culture bifurcation.
6 Originally published in the *Daily Telegraph*, 31 October 1998 (quoted here in Sagar, 2006: xi).

References

Abram, David, *The Spell of the Sensuous: Perception and Language in a More-than-Human World* (New York: Vintage Books, 1996).

Bentley, Paul, *The Poetry of Ted Hughes: Language, Illusion & Beyond* (London and New York: Longman, 1998).

Campbell, Joseph, *The Way of Animal Powers* (New York: A. van der Marck, 1983).

Carson, Rachel, 'Help Your Child to Wonder', *Woman's Home Companion* (July 1956), n.p. Republished posthumously as *The Sense of Wonder* (New York: Harper & Row, 1965).

Clark, Timothy, *The Cambridge Introduction to Literature and the Environment* (Cambridge: Cambridge University Press, 2011).

Cobb, Edith, 'The Ecology of Imagination in Childhood', *Daedalus*, 88:3 (Current Work and Controversies, Summer 1959), pp. 537–548.

Cushman, Keith, 'Hughes' Poetry for Children', in Keith Sagar (ed.), *The Achievement of Ted Hughes* (Manchester: Manchester University Press, 1983), pp. 239–256.

Dobrin, Sidney I. and Kenneth B. Kidd (eds), *Wild Things: Children's Culture and Ecocriticism* (Detroit, MI: Wayne State University Press, 2004).

Eliade, Mircea, *Shamanism: Archaic Techniques of Ecstasy* (Bollingen Series, 76. New York: Pantheon Books, 1964).

Gifford, Terry, 'Gods of Mud: Hughes and the Post-Pastoral', in Keith Sagar (ed.), *The Challenge of Ted Hughes* (Basingstoke and London: Macmillan, 1994), pp. 129–141.

Gifford, Terry (ed.), *Green Voices: Understanding Contemporary Nature Poetry* (Manchester: Manchester University Press, 1995; 2nd edition, Nottingham: Critical, Cultural and Communications Press, 2011).

Gowar, Mick, 'Poetry in the Making: Fifty Years Old', *Ted Hughes Society Journal*, 2 (Winter 2011), pp. 1–13. Web. 30 December 2011. www.thetedhughessociety.org.

Graves, Robert, *The White Goddess: A Historical Grammar of Poetic Myth* (London and Boston: Faber & Faber, 1948; New York: Farrar, Straus & Giroux, 1966).

Hughes, Ted, *The Hawk in the Rain* (London: Faber & Faber, 1957).

Hughes, Ted, *Lupercal* (London: Faber & Faber, 1960).

Hughes, Ted, *Poetry in the Making* (London: Faber & Faber, 1967).

Hughes, Ted, *The Iron Man* (London: Faber & Faber, 1968).

Hughes, Ted, *Remains of Elmet* (London: Faber & Faber, 1979).

Hughes, Ted, *Ffangs the Vampire Bat and the Kiss of Truth* (London: Faber & Faber, 1986).

Hughes, Ted, *Moortown Diary* (London: Faber & Faber, 1989).

Hughes, Ted, *Winter Pollen* (London: Faber & Faber, 1994).

Hughes, Ted, *Collected Plays for Children* (London: Faber & Faber, 2001).

Kazzer, Claas, 'Difficulties of a Bridegroom', in Bertrand Rougé (ed.), *Q/W/E/R/T/Y: Littératures et Civilisations des Pays Anglophones* (Pau: Publications de l'Université de Pau, 9 October 1999), pp. 187–201.

Lockwood, Michael, 'Ted Hughes: The Development of a Children's Poet', *Children's Literature in Education*, 40:4 (December 2009), pp. 296–305.

Louv, Richard, *Last Child in the Woods: Saving Our Children from Nature-Deficit Disorder* (Chapel Hill, NC: Alqonquin Press, 2005).
Melson, Gail F., *Why the Wild Things Are: Animals in the Lives of Children* (Cambridge: Harvard University Press, 2001).
Morin, Edgar, *Science Avec Conscience* (Paris: Fayard, 1982).
Moss, Stephen, *National Childhood Report* (National Trust, 2012). Web. 20 September 2016. www.nationaltrust.org.uk/documents/read-our-natural-childhood-report.pdf.
Paul, Lissa, 'Inside the Lurking-glass with Ted Hughes', *Signal: Approaches to Children's Books*, 49 (1986), pp. 52–63.
Paul, Lissa, 'The Return of the Iron Man', *The Horn Book Magazine*, 76:2 (March/April 2000), pp. 218–225.
Plumwood, Val, *Feminism and the Mastery of Nature* (London: Routledge, 1993).
Reddick, Yvonne, *Ted Hughes: Environmentalist and Ecopoet* (Basingstoke: Palgrave Macmillan, 2017).
Roberts, Neil, *Ted Hughes: A Literary Life* (Basingstoke: Palgrave Macmillan, 2006).
Robinson, Craig, *Ted Hughes as Shepherd of Being* (Basingstoke: Macmillan, 1989).
RSPB, *Connecting with Nature: Finding out How Connected to Nature the UK's Children Are* (2013). 20 September 2016. Web. www.rspb.org.uk/connectionmeasure.
Sagar, Keith, *Literature and the Crime Against Nature* (London: Chaucer Press, 2005).
Sagar, Keith (ed.), *The Laughter of Foxes* (Liverpool: Liverpool University Press, 2006).
Scigaj, Leonard M., *Ted Hughes* (Boston, MA: Twayne, 1991).
Shepard, Paul, *The Others: How Animals Made Us Human* (Washington, DC: Island Press, 1996).
Soper, Kate, *What is Nature? Culture, Politics and the Non-human* (Oxford: Blackwell, 1995).
Whitley, David, '"Imaginary Gardens with Real Toads in Them": Animals in Children's Poetry', in Morag Styles, Louise Joy and David Whitley (eds), *Poetry and Childhood* (Stoke-on-Trent and Sterling, USA: Trentham, 2011), pp. 187–194.
Wilson, Edward, *Biophilia* (Cambridge: Harvard University Press, 1984).
Wilson, E. O. and Stephen R. Kellert (eds), *The Biophilia Hypothesis* (Washington, DC: Island Press, Shearwater Books, 1993).

Part I

Speaking through the voice of nature

1 A life close to nature

The make-up of Hughes's imaginary self: childhood, landscape and memory

Edward James Hughes was born in astrological terms at 'solar midnight' at No. 1 Aspinall Street, Mytholmroyd, in the Calder Valley, West Yorkshire on 17 August 1930. According to his natal chart, like the stereotyped Leo, he was endowed with a strong sense of self. However, because of his 'powerful birth hour', paradoxically he required seclusion, as indicated by his ascendant: cancer, associated with home and family. He once wrote that his natal horoscope and astral influences had foretold his destiny and conflicts, making him 'fated to live more or less in the public eye, but as a fish does in air' (Middlebrook, 2004: 51).[1] Hughes was a very private person and hated being scrutinised by others. However, as fate would have it, the public tragedies and unfortunate events that he witnessed in his much publicised personal life often caused him to be the centre of public attention.

Literary critics and scholars have often regarded Hughes's esoteric side and genuine interest in astrology with light-hearted humour and somewhat uncomfortable bemusement. However, to the poet astrology *did* work; whether magic or science, it was not something he took light-heartedly and regarded it as an art. In his essay 'Superstitions' collected in *Winter Pollen* (1994) he presents his own defence, claiming that whether 'astrology works as an esoteric science, like advanced mathematics, as astrologers claim, or as an intuitive art, like throwing the bones, it doesn't matter, so long as it works' (*WP*: 51). Hughes was somewhat of an expert reader of horoscopes and applied astrology throughout his life, both by drawing on such symbols in his poetry and using it for important decision making, as in the case of publishing dates,[2] to the point that astrology became almost an obsession for the poet.

Astrology aside, Hughes's imaginary self was also deeply influenced by his native Yorkshire, and in particular the Calder Valley, as reflected in the internal landscape of Hughes, strongly related to his family history where he spent his first seven years and later returned throughout his life. In his collection of poems *Remains of Elmet* (1979)[3] Hughes celebrates his Celtic past and claims that the Calder Valley was originally the kingdom of Elmet, the last Celtic land to fall to the Anglo-Saxons. Like the Augustan poet Oliver Goldsmith's sentimental pastoral 'The Deserted Village' (1770), Hughes looks back at his origins and how Calder became 'the hardest-worked river in England' (*RE*: 8) only to return to witness the decline that had transformed a town that had once been the cradle of the textile industry into derelict mills, empty buildings and ghost memories. However, the poet, unlike Goldsmith, avoids sentimentalising the landscape of his childhood, where he claims to 'have watched the mills of the region and their attendant chapels die' (*RE*: 8). This is the geography of the spiritual make-up of Hughes's childhood that was grafted into his poetic landscape and imprinted as part of his vision of childhood. Hughes uses the metaphors of his most immediate childhood world and its setting to describe and develop his own feelings and shape his own poetic universe. However, the true importance of topography and of Hughes's childhood landscape is perhaps best stated by Hughes himself in an autobiographical prose piece about his early childhood entitled 'The Rock', commissioned by the British Broadcasting Corporation (BBC) in 1963, in which he described the effect of growing up in the Calder Valley. In it he describes Scout Rock, the large geographical feature which dominates the Mytholmroyd landscape and the underlying connection that existed between the poet and the landscape:

> This was the *memento mundi* over my birth: my spiritual midwife at the time and my godfather ever since – or one of my godfathers.… If a man's death is held in place by a stone, my birth was fastened into place by that rock, and for my first seven years it pressed its shape and various moods into my brain.
>
> (Hughes, 1963: 421)

The presence of Scout Rock in his childhood landscape was both totemic and mythological to Hughes to the point that it provided 'both the curtain and backdrop to existence' (1963: 421). Drawing on cultural geographer Yi-Fu Tuan's term *topophilia*, which can be defined as 'the affective bond between people and place or setting' (1974: 4), this early landscape was for Hughes, more than an 'affective bond', a barrier that both protected and 'sealed' (1963: 421) him from the outside world. Despite the apparent contradictions, it was to become a symbol of his identity and culture, a powerful and threatening presence with an 'evil eye' (Hughes, 1963: 422) that watched over him from his first day:

> It was not a frightening presence, it was a darkening presence, like an over evident cemetery. Living beneath it was like living in a house haunted by a disaster that nobody can quite believe ever happened, though it regularly upsets sleep.
>
> (Hughes, 1963: 422)

The geographical make-up of his childhood was to become the map for Hughes's own creative universe. By recalling the area, Hughes engages in the memory of its, landscape and the modes of awareness expressed through the sounds and signs that depict the magic of the place that he knew as a child. Throughout 'The Rock' Hughes recreates the journey and the inhospitable spirit of the craggy moors by 'storying' it from his memory as a young boy.

The essay illustrates the powerful link that exists between the memory of Hughes's childhood world constructed through landscape and experience, and Hughes's own imaginative universe, both of which form part of Hughes's art of storytelling. This can be seen in how Hughes emphasises the negative qualities reflected on him at the time he wrote it, as a kind of self-mythology. The underlying sense of topophilia also reveals the personal cost of commitment in Hughes's psychology and reads as a kind of 'religious awakening' in the light of what might now be called ecopsychology. As children's literature critic Peter Hollindale has suggested, the proposition of constructing one's self through memory, and the idea that our identity depends on our sense of personal continuity in time is expressed by storying our lives. Hughes frequently draws upon the memory of childhood and experience throughout his work for inspiration and images that shape his storytelling. However, experience and memory are both unstable elements since memories do not remain stable, for they 'are not constants: we revise and alter them to fit our present needs' (Hollindale, 1997: 71). This need to revise memories, more often than not explains the conflicting narratives and inconsistencies that frequently appear in Hughes's own writing, as Claas Kazzer has correctly pointed out 'Ted's stories were evading a reading for "truth" as a simple one-to-one correspondence. They contained historical and emotional truths, or "inner" and "outer" truths, if you will, to which he added the intrinsic truth of a good story' (2010: n.p.).

In the case of Hughes, the underlying bond between the writer and his early childhood landscape can be seen in the way that the poet frequently draws upon his childhood experiences for inspiration and metaphors to shape his poetry. There is a clear link between Hughes as a young boy absorbed in his own world and the memories that he has preserved and which have developed and been transformed into his writing. As Hughes himself said in an interview with Ekbert Faas for *London Magazine*, 'My first six years shaped everything' (Faas, 1971).[4] The juxtapositions between Hughes's inner and outer world together with his literary landscapes formed by real impressions as well as fictive visions and memories formed from his own Yorkshire childhood environment largely explain the thin line separating 'truth' and fiction.

The compelling significance that the moors had for Hughes, can be revealed not only in his childhood but as a presence that accompanied him throughout his life, as can be seen in *Remains of Elmet* (1979), where the poems depict the creative/destructive force of nature through a harsh landscape and its people, where 'The sunk mill-towns were cemeteries' (53) and where the only thing that flourishes is nature 'out of nothingness into nothingness' (110). That same 'nothingness' could be seen 'in the inane frozen-looking eyes of the sheep' (Hughes, 1963: 423), and the sense of mourning that pervades the entire landscape, as Hughes recalls as a young boy, and the lasting impression it had on him:

> I suppose in some ways it was eerie, and maybe even unpleasant. But everything in west Yorkshire is slightly unpleasant. Nothing ever quite escapes into happiness. The people are not detached enough from the stone, as if they were only half-born from the earth, and the graves are too near the surface. A disaster seems to hang around in the air there for a long time.
>
> (1963: 423)

Following this grim portrait, there is little doubt that the poet felt repelled by the industrial nature of his childhood landscape. Notwithstanding, it is important to point out that Hughes is not always an objective or reliable source since he often exaggerates the truth and indulges in self myth-making, especially in an autobiographical sense and when referring to his childhood. What is certain is the fact that his early experiences wandering the moors generated a sense of ambivalent commitment to the area. The moors became for Hughes, as for other local writers such as Emily Brontë, a kind of spiritual home, one which was both exhilarating and at the same time desolate. Moreover, as can be seen by the way in which Hughes depicts his environment, it was a way of connecting with his surroundings, and his immediate natural landscape; at the same time as he was developing a kind of bond with that surrounding landscape and the natural world that he would keep for the rest of his life.

It was of course on those same windswept rocky moors where Hughes, as a young boy, would often go hunting and shooting with his older brother, Gerald, who acted as his guide and protector. Those early adventures formed the bedrock of his imagination and experience, the essence of his childhood core. It is important to mention here that neither hunting nor shooting were typical pastimes for the working people of the Calder Valley, or their children. However, to Hughes and his brother Gerald they were a common pursuit to the point of becoming an obsession: one which would remain with him for his entire childhood and for a large part of his teenage years. As Hughes himself later confessed, up to the age of 17 or 18 shooting and fishing 'were pretty well my life, apart from books' (Heinz, 1995: 77). This also meant that as a young teenager he led two different lives:

> Up to twelve or thirteen I also played with my town friends every evening, a little gang, the innocent stuff of those days, kicking about the neighbourhood. But weekends I was off on my own. I had a double life.
>
> (Heinz, 1995: 77)

Gerald would also take his younger brother fishing and on camping trips in the Calder Valley, as Hughes recreates in the ghost story 'The Deadfall' (1994), partly autobiographical, at least in motifs if not in facts, published in *Difficulties of a Bridegroom* (1995), which curiously (or not?) coincided with the publication of his essay 'The Burnt Fox' in *Winter Pollen* (1994).[5] What makes them significant is that both accounts are concerned with the guilt at killing animal spirits.

Both his brother's mentorship and the landscape of his childhood created in Hughes a strong sense of alertness and the ability to learn to observe nature with the eye of a naturalist, and even more importantly a desire and fascination that developed and which would remain with Hughes all his life: his interconnectedness with nature. This underlying link can best be appreciated in his animal poems that express the celebration of animal life and the relationship between the energies and vitality of the human and non-human world. By identifying with nature and being in contact with the natural world Hughes came to think of his poems 'as a sort of animal', one he was able to 'capture' in words, as he put it in *Poetry in the Making* (1967: 15).

Hughes's interest in the natural world was however firmly rooted over 30 years before he wrote *Poetry in the Making*, in his desire to become familiar with his local environment by exploring nature and knowing the creatures that inhabited the landscapes of his childhood in depth. In 1938, however, that childhood landscape changed abruptly when, much to both young Hughes and his sister Olwyn's dismay, their parents decided to buy a newsagent's and tobacco shop in the mining town of Mexborough, South Yorkshire. As Steve Ely puts it 'there can be no doubt that Hughes was repelled by Mexborough's built-up and industrial nature and its associated pollution and blight' (2015: 183). Hughes himself later describes it as 'a mining town right in the thick of the coal and steel belt' (*LTH*: 623), next to the 'poisonous' river Don (*LTH*: 623). Gerald was 17 at the time and went to Devon and took up what, to Hughes, seemed to be an idyllic life as a gamekeeper. He then enlisted in the Royal Air Force (RAF), two years later when the Second World War broke out. Although Hughes continued to explore new farmlands and woodlands, his idyllic childhood dream had been shattered. Hughes worshipped his brother and his departure left an empty space not only in Hughes's life but also in his imaginative universe. Indeed, when his brother migrated to Australia after the war, Hughes spent a large part of his lifetime trying to persuade him to return home, living in hope and truly believing that one day they would finally be reunited. The loss of his fraternal figure and guide was partly taken up by the figure of his sister Olwyn, two years older than the poet, and reshaped by her interest in books, especially poetry.

The books that built Hughes

As a young child Hughes's home had few books, apart from the magazines and comics from his father's newspaper shop such as *The Shooting Times* and *The Gamekeeper* which he would avidly read, and a children's encyclopaedia that his mother gave to him, rich in Yeat's Celtic folktales (most probably Yeats's *Fairy and Folktales of the Irish Peasantry*, 1888).[6] Criticism has often overlooked the importance of the books that Hughes read in his childhood. These first books are of course the ones which formed the poet, having a strong impression and shaping his imagination whilst helping him to find his own form and style of writing. The imprint of those early reading experiences was fundamental. Not only in the influence of the style of writing but also in the themes and messages conceived by the child as a reader and which are also embedded in his writing. Hughes talks about his experience as a young reader and how this influenced his writing

> My writing grew mainly out of my reading. I specialized in fantastic happenings and gory adventures. I set my tales, generally North of the Great Divide, in the Wild West, in tropical jungles, and in Africa – places as far away as possible from the sooty town I lived in and the gloom of the War.
>
> (*WP*: 4)

Both the town where Hughes grew up and his father, William Henry Hughes, had suffered the horror of the First World War. Hughes's father, a veteran, rarely spoke of his experiences but 'would sometimes cry out at night in his sleep' (Roberts, 2006: 13). The silence of a psychologically wounded father, together with the experience of having lived the reality of wartime as a child growing up in the grim industrial town of Mexborough, must have surely conditioned Hughes own sense of his home town as being a gloomy, bleak and depressing place. That sense of bereavement is also recalled in his largely negative description of the West Yorkshire landscape where he grew up, so that his epic and archetypal imaginary worlds of 'fantastic happenings and gory adventures' (*WP*: 21) can be seen as a way of escapism into the fantasia of books amid new worlds of imagination, against the crude backdrop of the historical background and the landscape of his rural surroundings. Together with Yeats, amongst the first books that he read from the age of about 11 that largely influenced his own style and inspired him as a writer are two key texts that had a critical and formative influence on him: Henry Williamson's *Tarka the Otter* (1927) and Rudyard Kipling's *The Jungle Book* (1894).

Hughes would later become firm friends with the brilliant but tortured writer Henry Williamson, who lived locally in Devon and died in 1977. Further than the fact that Hughes read Williamson as a child, critics have generally refused to see any similarities between both writers, perhaps because of Williamson's affinity with fascism and the Hitler Youth Movement. However, below the surface and leaving political sympathies and differences behind, both shared important affinities that would mark their writing and approach to nature. Although they were both hunters, as ecologists they strived throughout their lives to protect nature and endure respect for the local countryside, the rivers and their inhabitants. Their writing projects a Darwinian view of nature, what the poet Alfred Tennyson in his masterpiece 'In Memoriam' (1850) called 'red in tooth and claw', since both depict violence and death as an integral part of the energy sustaining the cycle of life. Williamson, like Hughes, was also a self-taught naturalist and an acute observer of animals. In both cases their profound knowledge of ecology and nature suffuses their writing (particularly concerning the local habitat and region where they grew up, in Williamson's case North Devon and particularly the River Torridge) together with his descriptions of the countryside, its animal inhabitants and the surrounding landscape. Like Hughes, Williamson's poetic prose is alchemistic in its intensity, concerned with small details, since he presents an otter's-eye view of the river, its banks and wetlands. Both authors write convincing animal narratives and poems, void of sentimental visions of nature, and evoke empathy rather than sympathy in the reader towards the non-human, without over-anthropomorphising the animal characters that inhabit their universes. Williamson's *Tarka the Otter* (1927) was probably the book that most inspired Hughes as a boy. As Hughes himself said 'From eleven to thirteen I read *Tarka the Otter* over and over' (*WP*: 4). Hughes later said when describing reading *Tarka the Otter* that 'It entered into me and gave shape and words to my world, as no book has ever done since. I recognised even then, I suppose, that it was something of a holy book, a soul-book' (Lane, 1999: 152).

Another important early literary influence that shaped Hughes as a child was Kipling. Hughes was first inspired by *The Jungle Book* (1894), probably closely identifying as a young boy with the life of Mowgli in the wild. Likewise, Kipling's *Just So Stories* (1902) had an imprint in *How the Whale Became*, first published in 1963; however, it was the rhythm of Kipling's poetry that held a stronger grasp on the young poet. It was after a class reading of *The Jungle Book* at the age of 13 that Hughes discovered at the local town library Kipling's *Collected Verse* (1912) which he imitated when he began to write his first poems about fishing and shooting:

> Kipling's poems were a revelation. Something about them, hard to define, appealed to me, but what took hold of me were his rhythms. They got into my head and evidently affected me greatly. From that point, I began to write all my verse tales in Kipling's lockstep rhythms and resounding deadlock rhymes.
>
> (*WP*: 5)

Together with Kipling and Williamson, Hughes mentions the King James Bible as one of his main sources of imagery, 'with its very different rhythms', which he read 'looking for the high moments and magical sentences' (*WP*: 5). In 1957 he wrote to his sister Olwyn that the Bible was his 'favourite book' (*LTH*: 100). From an early age, Hughes held strong views about religion. In an interview with Ekbert Faas in *London Magazine* (January 1971), the poet explained how 'You spend a lifetime learning how to write verse when it's been clear from your earliest days that the greatest poetry in English is in the prose of the Bible' (10). For Hughes poetry had a spiritual dimension, similar to that of religion, as he wrote in a letter to Bishop Ross Hook on 10 November 1982 'poetry seems to point, on one bearing or another, towards a spiritual life of sorts, or perhaps towards something that is a substitute for the spiritual life as it is understood by the Church' (*LTH*: 457). The question of whether Hughes's poetry serves the same function as religion requires more careful elucidation. As will be discussed later, the shamanic sense of Hughes's poetry is not necessarily religious in a sacred or biblical sense, but more in a primitive, deep-seated and spiritual sense. Hughes wrote using his in-depth knowledge of myths and symbols to restore, to liberate and to heal.

The teachers who inspired him at Mexborough Grammar School were also quick to notice his talent and encourage his writing. It was in fact John Fisher, his charismatic English teacher, who introduced the young Hughes to the world of Eliot and Hopkins, whilst Miss McLeod, the headmistress, and Pauline Mayne,[7] who held John Fisher's post whilst he was in the armed forces, remarked on his ability to write poetry. It was also at Mexborough that Hughes took his first steps on stage and developed his ongoing interest in drama, although his first real connection to drama came through encouragement from Olwyn, who was at the time his literary mentor, to read Shakespeare. From then on Hughes read and reread Shakespeare throughout his life. Between 1949–1951, whilst Hughes was doing his National Service in the RAF, stationed at an isolated radar station on the Yorkshire moors, he 'had nothing to do but read and reread Shakespeare and watched the grass grow' (Sagar, 1975: 8).

Together with Shakespeare and the work of the Romantic poets, one of the most influential books he read during those pre-university years was Robert Graves's *The White Goddess* (1948), which he had been given as a personal leaving gift by Fisher. Critics have long recognised the influence of Graves's book, written as an investigation of poetic myths, in Hughes's poetry. According to Gifford and Roberts: 'The myth that has most consistently inspired Hughes is the subject of Robert Graves's *The White Goddess:* the nature goddess in her three aspects of maiden, mother and crone (Persephone, Demeter and Hecate in the best-known version)' (1981: 19). The full extent of Hughes's debt to Graves becomes clearest in his colossal critical study *Shakespeare and the Goddess of Complete Being* (1992).

In addition to Graves's book being idiosyncratic, it is also somewhat obscure to read and to categorise given its dense train of thought. In it, Graves identifies poetry with religion and claims that the earliest religion of mankind is the religion of the goddess. Not only did the book have a profound influence on Hughes's writing and work, but also on his creative development and poetic mode of thought. The poet found in Graves a voice, committed to the goddess and her archetypes, that was the total embodiment of his own poetic beliefs. In a letter he wrote to Graves in 1967 he acknowledges that the book was 'the *chief holy book* of my poetic conscience' (*LTH*: 273) and that he felt a slight resentment on his first reading of *The White Goddess*, since Graves had taken possession of what he considered to be his own 'secret path' (*LTH*: 679). Hughes's key interest lies in the cycle of nature and the forces and elements that control the energies behind this cycle. It was through the figure of the goddess and the power of myth that Hughes explored in his poetry the primal energies of nature and the creative and destructive forces of the universe. Hughes's poetry stresses these dualities as well as the sense of absolute otherness of nature, and conveys his concept of the relationship between these energies and that of 'Western man'[8] (*WP*: 129) who has become alienated from the natural world.

Hughes's poetic voice had of course already revealed in his early childhood a constant fascination with the natural world, and especially wild animals, thanks to his adventures with his brother and his interests in fishing, shooting and trapping animals. However, as Middlebrook points out, 'This is what will eventually separate him from Gerald once and for all. Hughes will transfer the skills of the hunter and fisherman into the enterprise of reading and writing' (2004: 76). This clear change of attitude in the poet's ideas related to the relationships portrayed between human/animal interaction can be observed when Hughes leaves childhood and enters adulthood.

His perceptiveness and the balance between imagination and observation are in many ways what made his poetic voice so perfect and what gave him a lifelong appreciation of poetry. However, at the age of 16 and at the end of his school days the poet's journey was only just beginning.

From Pembroke to 'The Burnt Fox'

In 1948 Hughes won a place at Pembroke College, Cambridge, to read English. Writing to his sister Olwyn in 1952, whilst in his first year, he comments on his rather dull life, and his routine:

> Sometimes I think Cambridge wonderful, at others a ditch full of clear cold water where all the frogs have died. It is a bird without feathers; a purse without money; an old dry apple, or the gutters run pure claret … I get up at 6, and read a Shakespeare play before 9, and sometimes half an hour's Chaucer as well. Then I range all day among readings and writings.
>
> (*LTH*: 12)

It is clear that Hughes felt a strong aversion to the academic study of literature at Cambridge. Like many, Hughes had hoped that English would improve his potential as a true writer but, by the end of his second year he realised with disappointment that his hopes had been shattered. After completing his first two years of studying English literature Hughes found the method of detailed literary analysis too limited and switched to archaeology and anthropology as academic areas more suited to his deepest interests, coinciding with his discovery of the literature of shamanism, and his own understanding from reading Mircea Eliade and Carl Jung. Bearing in mind that Hughes had already read Robert Graves's *White Goddess* before starting his final year studies, Hughes's background and own interests were also well suited to this change.

The catalyst of his decision to change studies however was his renowned burnt fox dream story, which would inspire his poem 'The Thought Fox' (Hughes 1957). The best account of this is given by Hughes himself in *Winter Pollen* (1994). In it the poet describes how one evening he tries in vain to find the words to complete his last essay for his weekly supervision. Exhausted the young poet decides to give up and surrender to sleep, when the figure of a fox appears before him, standing upright on its hind legs with its body charred:

> Its eyes, which were level with mine where I sat, dazzled with the intensity of the pain. It came up until it stood beside me. Then it spread its hand – a human hand as I now saw, but burned and bleeding like the rest of him – flat palm down on the blank space of my page. At the same time it said: 'Stop this – you are destroying us.'
>
> (*WP*: 8–9)

In retrospect, for Hughes the dream was more than a sign, a shamanic experience. From it Hughes later explicitly wrote 'The Thought Fox', the poem he identified as the first of his adult poetic career, a poem, as he puts it, about 'both a fox and a spirit' (*WP*: 14). Moreover, there are significant references throughout Hughes's work to suggest that the fox was an important totemic figure for Hughes, the first of many which would accompany him throughout his life (another clear example of the importance of the fox as a totemic figure can be seen through the fox cub and the ivory fox that appear in the story 'The Deadfall', 1995). Neil Roberts describes 'The Burnt Fox' as the 'founding myth of Hughes's poetic career' (2006: 12) and almost 'a category of its own' (2006: 28). Despite the outstanding amount of critical attention and analysis of the poem and indeed on the fox figure and its importance for Hughes, it appears that critics have failed surprisingly to ask the simple question of 'Why a fox? (as say opposed to another animal?)' – and link the fox to a representation of his early hunting days, given its status as a local wild animal.[9] Unlike Hughes's other totemic figures (including the crow, the hawk and the jaguar that frequently appear in his work) the fox as a local wild mammal would appear to be closer to his childhood memories and experiences and can be seen as playing a double role as both hunter and hunted.

The change of studies brought fresh air to Hughes, as can be seen in a letter he wrote to his brother Gerald, in November 1953. In it the poet comments on his 'new life style' which entailed eating six eggs whipped in a pint of hot milk and sugar as part of his new 'power scheme' and roasting meat over a fire in his room in the middle of the night:

> I go to bed about 11, and get up at 2:30 in the morning. Then I work, and sing, and spring about amusing myself and draw grotesque figures on my walls until about 7, then I go to bed until 9:30. By this means I get two first sleeps, and feel fresh as a flower. The second (7–9:30) sleep is always full of the most marvellous dreams, and this is the best part of my day. This is like leading two lives.
>
> (*LTH*: 22)

The bizarre drawings mentioned by Hughes in his room also appear in the book *Memories of Ted Hughes: 1952–1963* (2010), written by Daniel Huws, a close friend of Hughes at Cambridge. In it he tells the comical story of the life size pumas and other creatures which adorned his walls at Pembroke:

> The story went that these had been reported to authority and that Mr Camps, the senior tutor, had come to see them for himself and to reprimand Ted, that Ted persuaded Mr Camps to lie down on the floor in order to appreciate them better.
>
> (Huws, 2010: 16)

Hughes's imagination drew on many different sources in his young adult years, including his 'sacred canon' formed in the case of poetry by the Romantic poets Blake, Wordsworth, Coleridge and Keats, and then later Hopkins and Eliot. In addition to Shakespeare and Chaucer, another author who Hughes admired and drew on, was Jonathan Swift. Writing to Olwyn in 1952, he states 'Swift's excellence is a talent for clarity simplicity and power. His inimitable peculiarities are in choice of subject, or tone, but his writing is the bed-rock from which every writer must start' (*LTH*: 20).[10]

Given the fact that Hughes's literary canon had then already been more or less shaped by the time he went to university, and taking into account that he had already read most of the authors of the Cambridge English canon of the time, it seems logical that Hughes found little inspiration in his English studies and felt that intellectually they had little to offer him. Together with his early readings of folklore, mythology, Robert Graves's *White Goddess* (1948) and Jung, Hughes's literary canon was strongly reinforced by his studies in Archaeology and Anthropology and his encounter with new authors including Charles Darwin, Claude Lévi-Strauss and Sigmund Freud. However, perhaps the two works which most influenced his writing and thought at that time were Joseph Campbell's *The Hero with a Thousand Faces* (1949) and James Frazer's monumental study in folklore, magic and religion, *The Golden Bough* (1890), both of which shaped his conception of the imagination as a creative and regenerative faculty alongside his extensive reading in Zen, Sufisim, Taoism, Eastern European poetry, mythology, the Cabbalah and of course his interest in Shamanism.

Shaman and shepherd: healing the wounds

Hughes reviewed Mircea Eliade's book *Shamanism* in 1964 and described shamanism not as a religion, but a 'technique for moving in a state of ecstasy amongst the various spiritual realms, and for generally dealing with souls and spirits, in a practical way, in some practical crisis. It flourishes alongside and within the prevailing religion' (*WP*: 56). As Neil Roberts underlines

> What particularly excited Hughes about ME's book, however, was not the idea of practising as a shaman in the traditional sense, but that the equivalent of the shaman in modern society is the poet, or a certain kind of poet with whom he identified himself.
>
> (2006: 110)

For Hughes shamanism was not only a spiritual revelation, but a way of explaining and incorporating everything that he had read and identified with up to then: his hunting experience together with fishing, his relation with the wild, folklore and mythology, and of course his attitude towards the natural world. In relation with his animal-centred shamanism, Chinese Hughes scholar Chen Hong has analysed how animal totems in Hughes animal poems function as critiques of human life and states that

> underneath them all, what he found was a deeper connection between animal life and the divine world, a world that animals have always been living in and that humans are separated from, a world that is sometimes termed 'the animal/spiritual consciousness'.
>
> (2011: 40)

In a letter to Moelwyn Merchant, Hughes explains his ideas about shamanism and how he understood shamanism as a divine state:

> I made the association, somehow, between the world of animals, which is excluded by culture, & persecuted (killed & eaten) & the 'real thing' in human beings – the part which our own culture tortures, i.e. sacrifices, crucifies. I identified, you see, the sacrificed God, the divine self which has to die to come into life, with the whole animal & vegetable kingdom, which culture tortures & destroys.
>
> (*LTH*: 460)

Hughes's sense of 'the divine state' can also be read here in the light of Swift and especially in the last part of *Gulliver's Travels*, where in part four the Houyhnhnms are depicted by Swift as 'divine' noble vegetarian horse-like creatures who rule the savage, meat-eating man-like Yahoos (Gadd *et al.*, 2014). The 'savage indignation' behind Swift's satires stems from his resentment of man's corruption and evil nature and the difference between what man could be *if* he rose to the full height of humanity. Hughes 'divine animal state' would appear to be very close to that of Swift's ideal. A human being, like Gulliver, does not have the same naked rational beauty as the horses, and throughout history, pushed by our irrational thoughts and unfounded acts we have enslaved others and become, like Gulliver, simply subequine primates. Once humanity has lost that divine state, for Hughes shamanism is the only way of regaining that alienated consciousness and recuperating the lost animal/spiritual state.

Following this clear change of attitude in the poet's ideas related to the relationships portrayed between human/animal interaction, Hughes was also becoming aware of his own destructive mode of engagement with nature and more distanced with his activities as a hunter by becoming more attuned with the natural world by reversing the roles of huntsman and victim and identifying instead with the hunted animal. Hughes's own writings and especially his environmental work, from the 1970s on, also suggest a conflict with this aspect together with a change of attitude towards the act of hunting resulting from his own engagement with the lives of animals.

Although Hughes did give up hunting he continued to fish throughout his life, and saw it as an activity that restored 'the lost paradise of atonement with the natural world and with the primal roots of one's own psyche' (Sagar, 2006: 147). Given that fishing is also a form of hunting, certain critics (Roberts, 2006; Sagar, 2006) have correctly regarded this attitude as being ambiguous and identified it as a contradiction.

In an interview that Hughes gave to Thomas Pero in *Wild Steelhead & Salmon* (published in 1999), shortly before his death, the poet talked with great passion about his fascination with fishing, salmon and the life of rivers. In it Hughes claims that if fishing were abolished he would have to leave England and says that 'I would never stop fishing, because I do not want to lose what goes with fishing ... this last connection. To this whole – to everything. The stuff of the Earth. The whole of life' (Pero, 1999: 55). Hughes would appear to be referring here to the fact that to him fishing was a way of 'reconnecting' himself with the cycle of life. To Hughes fishing and hunting were a way of restoring the primitive side to man, a way of 'reconnecting the circuits' and restoring the biological inheritance of our primitive soul 'the feeling of something absolutely sacred and unspoiled' (Pero, 1999: 56).

Indeed, the ethical debate and question behind this – whether or not a human being has the right to disturb another being by intruding in its habitat and taking that life – remains problematic. For this reason, it has seemed wise to go no further than to acknowledge Hughes's somewhat contradictory position on the whole subject of hunting, although in his works for children he is consistently empathetic towards the animals hunted.

This concern and split in Hughes's way of thinking is also closely linked to Carl Jung's philosophy of 'reconnecting with nature' in order to awaken our instincts as a way of 'finding' ourselves. Hughes's Jungian desire and his sense of 'reconnecting' correspond to his wish to restore the balance of his psyche and heal the wounds of the crime against nature that he has committed. By distancing himself with the figure of the hunter and identifying himself with the creatures of the natural world Hughes becomes nature's alchemist and in this way is able to undertake his role as the shaman who is led by the animal world.

Mircea Eliade, in *Shamanism* (1964), explains how 'the shaman is the great specialist in the human soul; he alone "sees" it, for he knows its "form and its destiny"' (8). As Eliade notes, there is a 'mystical solidarity' that exists between humans and animals so that

> each time a shaman succeeds in sharing the animal mode of being, he in a manner re-establishes the situation that existed in *illo tempore*, in mythical times, when the divorce between man and the animal world had not yet occurred.
>
> (1964: 94)

It seems obvious then that Hughes thinks of his dedication to writing in terms of shamanism. The poet as a shaman is a spiritual seeker on a quest journey. However, it is important to point out that Hughes was not simply on a *personal* quest of self-realisation, but his purpose was also to help others, although like the shaman, in helping others and through his writing and work he also received help himself in the form of redemption or healing.

Returning to Jung's philosophy of 'reconnecting ourselves with nature' and in relation with Hughes's notion of the significance of healing for human civilisation, something that major critical studies of Hughes appear to have surprisingly failed to notice is that Hughes's concern with our disengagement and loss of connection with the natural world is also very similar to that of Jung. Both Hughes's and Jung's close connection to nature can be attributed to the fact that as children they both grew up in the countryside and felt closely connected to the natural world.

Hughes's own 'voice of nature' portrayed in his children's writing does not depict a sentimental idealised world of innocence or anthropomorphic representations of how humans perceive animals but, as we shall see, identifies instead with the otherness of nature. Both Martin Heidegger and deep ecologists such as Arne Naess have illustrated how the anthropocentric development of Western philosophy has affected human culture and encouraged humans to live as if they were divorced from their environmental surroundings. It is this separation that Hughes attempts to heal in his children's writing, by advocating a sense of connection and not separation between humans and nature.

Throughout his life Hughes aspired to teach children to respect the natural world through a positive connection to nature. Hughes's concern that children were becoming increasingly disconnected from the natural world is similar to that voiced more recently by authors such as Richard Louv, who uses the term 'nature-deficit' to refer to today's generation of children (2005), or Jay Griffiths (2014). Both refer to the lost paradise of childhood freedom in nature and claim that today's generation of children are growing up detached from physical enjoyment with the earth.

Jung's concern for this same loss of connection with nature can also be discerned in many of his writings, in which he emphasises the importance of remembering what it is in the history of humankind that we have lost. Both Hughes and Jung ask similar questions in their writing: How did our loss of connection with the natural world come about? What consequences does this entail? And more importantly how might this rupture be healed? Similarly, both address the individual within a culture that is suffering and in need of help.

In 1928, Jung made the following provocative remark to a dream seminar 'Matter in the wrong place is dirt. People got dirty through too much civilization. Whenever we touch nature, we get clean' (Sabini, 2002: 1). Jung lived to see the emergence of the same techno-industrial age that Hughes condemns in his writing. The Swiss psychoanalyst sets our alienation with nature in the context of the development of consciousness over the last century, and felt, like Hughes, that Western consciousness had developed through science and technology, whilst leaving behind art, and forsaking our own spiritual growth. Jung also warned of the dangers of going against nature and claimed that 'by focusing almost singularly on developments in the outer physical world, what we have neglected is ourselves, our own inner nature' (Sabini, 2002: 11). Both Hughes and Jung convey a similar deep-seated respect for nature as a 'divinity' and identify our loss of connection with nature not as a psychological problem but a spiritual one. As Jung explained in his acclaimed *Letters I and II* (1975),

> Man ... is a top animal, exiled on a tiny speck of a planet in the Milky Way. That is the reason why he does not know himself; he is cosmically isolated. He can only state with certainty that he is no monkey, no bird, no fish, and no tree. But what he positively is, remains obscure.
>
> (cited in Sabini, 2002: 13)

In a similar way to Hughes, and in line with Darwinian evolutionary theory, Jung reminds us that humanity has forgotten that human beings are also primates, and it is through this degeneration of instincts that humans have become dissociated from nature and in need of healing.

Given the imprint of Jung's work in Hughes, it would seem that there is a place for Jung in relation to understanding Hughes's own environmental concerns, especially concerning the poet's sense of ecological responsibility, his sense of healing and the interconnectedness with animals. Both Jung and Hughes were ahead of their time and tried to voice their concern for environmental degradation, and alert people of the danger of manipulating natural resources by warning of the consequences not only in relation to the natural world but by showing that the contamination of our air, water and soil is parallel to that of diseases of blood and tissue in animals and people. As Hughes tellingly wrote 'What you throw into the ditch, appears in your child's cup.'[11]

Hughes's environmental consciousness is clearly manifested in his notion of responsibility for the management of nature and the underlying importance that he gives in his children's work to environmental education. For it is here where Hughes speaks as a steward of nature, and by assuming the role of a shaman and shepherd, and by *becoming* nature he also advocates in his children's writing empathy towards the natural world.

Notes

1 Undated letter Ted Hughes to Leonard Clark, postmark 1974, Berg Collection, New York, Public Library, New York. Quoted here in Middlebrook (2004).

2 As, for example, in the case of *Wodwo* (1967), when he urged his editor Charles Monteith to ensure it was published when he believed the stars were 'right' (*LTH*: 266).

3 *Remains of Elmet* (1979) was published by Faber and Faber and later republished as *Elmet* (1994) with new poems and photographs.

4 The interview was first published under the title 'Ted Hughes's Crow' in *London Magazine*, January 1971 and was reprinted in Faas (1980: 83–97).

5 'The Burnt Fox' (*WP*: 8–9) recounts a dream Hughes had whilst at university which inspired him to change his studies from English to Anthropology and which inspired his poem 'The Thought Fox', collected in *The Hawk in the Rain* (1957).

6 Hughes was to become an avid reader of Yeats, and would later read Yeats's *The Wanderings of Ossian* (1889) which he obtained from the library at Mexborough Grammar School.

7 Hughes corresponded with his former teacher, Pauline Mayne, for many years and often sent her copies of his poetry books. The British Library holds several letters that Mayne sent to Hughes. In one dated 24 April 1995, she thanks Hughes for his collection of poems (this would have been *New Selected Poems*, published March 1995) and remarks that,

> I've been involved reading some of the poems I've not seen before. I like 'Anniversary' very much. I think? But I'm still digesting. Are you not sometimes shocked by the amount you must have thought, felt, understood, to write so many poems? You're not old enough.
>
> (Add MS 88918/35/23)

8 'Western man' (*WP*: 129) is Hughes's term. Hughes uses the generic sense of man here, meaning 'human beings', in a way which many ecocritics would consider dated and sexist.

9 It is interesting to point out that according to Celtic legends (which Hughes would have been familiar with), the fox is seen to be a guide and since the fox's habitat was situated in the woods, the Celts relied on the fox as their guide in the spirit world.

10 The influence of Swift's style and irony can be seen clearest in Hughes's children's writing, especially in the case of his animal fables.

11 The quote comes from a notebook at the British Library containing various drafts of *Season Songs* (Add MS 88918/1/55). Amongst the drafts lies a speech Hughes gave as Poet Laureate about farming and the environment in which he mentions the dying of fish in his childhood rivers, Carson's *Silent Spring* (1962), DDT, cancer and ends with the above words.

Archive material

Add MS 88918/35/23. Personal and Writers' Correspondence Mayne–Muldoon (The British Library, Western Manuscripts, London, 7 August 1958–28 August 1998).

Add MS 88918/1/55. *Season Songs* Drafts (The British Library, Western Manuscripts, London, 18 August 1974–1979).

References

Carson, Rachel, *Silent Spring* (Boston, MA: Houghton Mifflin, 1962; London: Penguin, 1999).

Eliade, Mircea, *Shamanism* (London: Routledge, 1964).

Ely, Steve, *Ted Hughes's South Yorkshire: Made in Mexborough* (Basingstoke: Palgrave Macmillan, 2015).

Faas, Ekbert, 'Ted Hughes and Crow: An Interview', *London Magazine*, 10 (10 January 1971), pp. 5–20.

Faas, Ekbert, *Ted Hughes: The Unaccommodated Universe* (Santa Barbara, CA: Black Sparrow Press, 1980).

Gadd, Ian, Terry Gifford and Lorraine Kerslake, 'A Note on Ted Hughes and Jonathan Swift', *Ted Hughes Society Journal*, 4:1 (2014), pp. 85–88.

Gifford, Terry and Neil Roberts, *Ted Hughes: A Critical Study* (London: Faber & Faber, 1981).

Graves, Robert, *The White Goddess: A Historical Grammar of Poetic Myth* (London and Boston, MA: Faber & Faber, 1948; New York: Farrar, Straus & Giroux, 1966).

Griffiths, Jay, *A Country Called Childhood: Children and the Exuberant World* (Berkeley, CA: Counterpoint, 2014).

Heinz, Drue, 'Ted Hughes: The Art of Poetry LXXI', *Paris Review*, 134 (Spring 1995), pp. 55–94.

Hollindale, Peter, *Signs of Childness in Children's Books* (South Woodchester, Gloucestershire: Thimble Press, 1997).

Hong, Chen, 'Hughes and Animals', in Terry Gifford (ed.), *The Cambridge Companion to Ted Hughes* (Cambridge: Cambridge University Press, 2011), pp. 40–52.

Hughes, Ted, *The Hawk in the Rain* (London: Faber & Faber, 1957).

Hughes, Ted, *How the Whale Became* (London: Faber & Faber, 1963).

Hughes, Ted, 'The Rock', *The Listener* (19 September 1963), pp. 421–423.

Hughes, Ted, *Poetry in the Making* (London: Faber & Faber, 1967).

Hughes, Ted, *Wodwo* (London: Faber & Faber, 1967).

Hughes, Ted, *Remains of Elmet* (London: Faber & Faber, 1979).

Hughes, Ted, *Shakespeare and the Goddess of Complete Being* (London: Faber & Faber, 1992).

Hughes, Ted, *Elmet* (London: Faber & Faber, 1994).

Hughes, Ted, *Winter Pollen* (London: Faber & Faber, 1994).

Hughes, Ted, *New Selected Poems 1957–1994* (London: Faber & Faber, 1995).

Hughes, Ted, *Difficulties of a Bridegroom* (London: Faber & Faber, 1995).

Huws, Daniel, *Memories of Ted Hughes: 1952–1963* (Nottingham: Five Leaves Publications, 2010).

Kazzer, Claas, 'What is the Truth? Ted Hughes, Childhood, Memories and Stories', 2010. Web. 24 March 2012 www.ted-hughes.info/criticism/online-articles/kazzer-claas/what-is-the-truth.html.

Lane, John, *In Praise of Devon: A Guide to Its People, Places and Character* (Totnes: Green Books, 1999).

Louv, Richard, *Last Child in the Woods: Saving Our Children from Nature-Deficit Disorder* (Chapel Hill, NC: Alqonquin Press, 2005).

Middlebrook, Diane, *Her Husband* (London: Little, Brown, 2004).

Pero, Thomas, 'So Quickly It's Over', *Wild Steelhead and Salmon*, 5:2 (Winter 1999), pp. 50–57.

Reid, Christopher (ed.), *Letters of Ted Hughes* (London, Faber & Faber, 2007).

Roberts, Neil, *Ted Hughes: A Literary Life* (Basingstoke: Palgrave Macmillan, 2006).

Sabini, Meredith (ed.), *The Earth Has a Soul: The Nature Writings of C.G. Jung* (Berkeley: North Atlantic Books, 2002).

Sagar, Keith (ed.), *The Art of Ted Hughes* (Cambridge: Cambridge University Press, 1975).

Sagar, Keith (ed.), *The Laughter of Foxes* (Liverpool: Liverpool University Press, 2006).

Tuan, Yi-Fu, *Topophilia. A Study of Environmental Perception, Attitudes, and Values* (Englewood Cliffs, NJ: Prentice Hall, 1974).

2 Reconnecting with nature

Instilling a sense of wonder

It is no accident that the practice of ecological literary criticism appeared shortly after the emergence of the modern environmental movement in the 1960s, following which environmental criticism began to take shape in the second half of the twentieth century, in response to perceptions of threats not only to wildlife but also to human health.

Hughes himself was deeply aware of the public consciousness of environmental problems and the social change that was beginning to take place in the 1960s and 1970s. In a letter dated 22 August 1957, only a couple of months after having moved with Sylvia Plath to her home in Massachusetts, Ted Hughes wrote to his sister Olwyn criticising the 'American way of life' together with its excess packaging and excessive food miles:

> What a place America is. Everything is in cellophane. Everything is 10,000 miles from where it was plucked or made. The bread is in cellophane that is covered with such slogans as de-crapularised, re-energised, multicramulated, bleached, double-bleached, rebrowned, unsanforised, guaranteed no blaspheming.
>
> (*LTH*: 106)

To Hughes, such indications were somewhat alarming whilst being part and parcel of what was to come. Almost a year later, in a letter to his brother Gerald, Hughes comments that 'Our year in Northampton is almost over, thank God. Of all plastic cellophane wrapped places on earth, this is the prototype' (*LTH*: 131). Whilst everyone else was referring to the 'progress' behind the modernisation of retail sectors in the American food industry Hughes seemed aware that it is often small changes that carry the potential for harm through everyday products that surround us. Already as a young man, he was conscious of the degradation of what we now call the environment, and the social change that was beginning to take place.

Hughes's version of America anticipates the fall from grace for America's suburbia that was picked up by Carson in *Silent Spring* (1962), often considered as 'the founding text of modern environmentalism' (Garrard, 2011: 2), and which, in Hughes's words, 'revealed the whole of America as a poisoned land' (Gifford, 1995: 131).

Prompted by a letter in 1958 from her friend Olga Owens, a Boston newspaper reporter, who had written to Carson to complain that pesticides were wiping out birds, she was to spend the next four years studying scientific studies and reports, and collecting material. Written as a study on the toxic effects of residues of industrial and agricultural chemicals in animal and human bodies, its main claim was to bring attention to the widespread use of pesticides and their effects on the environment. The first chapter 'A Fable for Tomorrow' opens revealing a peaceful fairy tale:

> There was once a town in the heart of America where all life seemed to live in harmony with its surroundings. The town lay in the midst of a checkerboard of prosperous farms, with fields of grain and hillsides of orchards where, in spring, white clouds of bloom drifted above the green fields.
>
> (Carson, 1962: 1)

Carson's pastoral portrait of nature shows a harmonious picture of humanity coexisting blissfully with nature. However, this soon turns into a fable of doom envisioned through a nightmarish picture of destruction and catastrophe when

> a strange blight crept over the area and everything began to change. Some evil spell had settled on the community: mysterious maladies swept the flocks of chickens; the cattle and sheep sickened and died. Everywhere was a shadow of death.
>
> (Carson, 1962: 2)[1]

Alerting us to the dangers of manipulating natural resources and destroying the environment was Carson's life-long concern. Translated into over 30 languages, *Silent Spring* was hailed by scientists whilst being condemned by chemical companies. Sadly, half a century ago, few people believed that humans could harm nature. It was necessary for the popularisation of the notion of ecology and for people like Rachel Carson to demonstrate the horrific effects that a single chemical, like DDT, could have on the whole food chain. As Carson's biographer Linda Lear puts it:

> I don't think Rachel should be or would want to be credited with starting the environmental movement or banning pesticides. I think what she was hoping to do is raise the American consciousness about the natural world and our interconnection to it, instead of thinking we can control nature.
>
> (2012: n.p.)

Years before the term 'ecofeminism' was coined, Carson embodied the movement through her work and writings, endorsing the notion that whilst humans are but a small part of nature, they hold the ability to alter it.

Like Carson, Hughes was also aware that you could not harm a part of nature without harming the whole. Collected in *Winter Pollen* is Hughes's statement of Carson's poetic apocalyptic parable that 'came as an absolute shock' (129) and which should have been a warning to change the attitudes of human beings. However, as Hughes notes 'After ten years, the evidence of that book has not reached the average gardener.... The colossal mass of evidence simply has not been marshalled and sent to the one front that counts: the ear of the public' (*WP*: 129). Hughes was, of course, profoundly influenced by Carson's work, however it is important to point out that both Plath and himself were already familiar with her writing and had read several of her books whilst in America years before *Silent Spring* had even been published. As can be seen from a letter he wrote to Aurelia Plath in July 1958:

> I brought a load of books out of the library. Two of them by Rachel Carson, books about life in the sea which are wonderful. Sylvia is in raptures. I started one yesterday and couldn't put it down till I had finished it completely.
>
> (*LTH*: 127)[2]

Plath also wrote to her brother, Warren, on 5 July, 1958 claiming to have become 'more and more desirous of being an amateur naturalist', whilst reading 'the delightful book *The Sea Around Us*, by Rachel Carson. Ted's reading her *Under the Sea-Wind*, which he says is also fine' (1976: 345).

Carson's influence can be traced throughout Hughes's writing for children, especially in his novels *The Iron Man* (1968) and *The Iron Woman* (1993) where Hughes draws strongly on Carson's seminal work. However, it was Carson's last book, published in 1965, just a year after her death, that left an even greater legacy than the immediate impact of *Silent Spring*. For it is in *The Sense of Wonder* (1965)[3] where Carson offers a solution for our ecological crisis by revealing that the path to building an ecological consciousness lies in responding to the awe of nature expressed by children (Kerslake, 2016). Raising environmental awareness and instilling in the reader a sense of connection and a deeper understanding of the natural world was what Hughes thought all poets should do. However, from the very beginning of his career Hughes went a step further, by making his environmental thinking public, and by taking on a more active and 'practical' role than Carson.

It was shortly after reading Carson that Hughes became concerned with the effects of ocean pollution on marine and human life and the impact and scale of environmental hazards it caused. In a radio interview for the British Broadcasting Corporation (BBC) in 1992 Nigel Forde asked Hughes whether his interest in ecological and conservation issues had changed the way he looked at landscape, animals and poetry. The poet replied that it had and dates his 'revelation of what was happening' (in Gifford, 1995: 131)[4] back to 1959, whilst in the USA and shortly after reading Carson. It was then, after fishing on Cape Cod and reading an article about the commercial dumping of atomic waste in *The Nation*, that

Hughes first became concerned with the effects of industrial pollution whilst discovering that the mackerel there were radioactive. It was following this 'revelation' that Hughes became concerned with the effects of pollution on all forms of life as well as the impact and scale of environmental hazards it caused.

Although Hughes environmental consciousness was triggered by his experience in the USA, the sense of *topophilia* in connection with Hughes's affective ties with the environment related to his own childhood experiences fishing in the local rivers of his Yorkshire upbringing should not be overlooked. When Terry Gifford asked Hughes to tell him the story of his 'greening' Hughes explained that:

> From my earliest days I was hooked on fish – but I lived by a river, the West Yorks Calder, that had no life in it at all. Straight industrial effluent. And the Don, in South Yorks, which I expect you've glanced into, was worse. So my greening began you could say with everything that lay about me in my infancy. What focused my ideas maybe was just that devastation.
>
> (Gifford, 1995: 132)[5]

After witnessing the decline in trout and salmon in the Devon rivers of the Taw and the Torridge in the mid-1980s Hughes wrote a campaign statement for a public enquiry to clean up the River Torridge. As well as being an active campaigner concerned about the water quality in rivers, he also sat on a committee for the National Rivers Authority and helped set up the Westcountry Rivers Trust in 1993. This devastation caused Hughes to not only observe the changes taking place around him locally in nature, but it also encouraged him to become actively involved in alerting people to the consequences of manipulating natural resources and destroying our environment, having witnessed their effects, and speaking in this sense as a steward for nature throughout his writing.

Hughes's greening and the ecological crisis

Environmentalists' claims like those of Carson laid the path to what can be considered as a 'literary' or 'cultural' analysis of the consequences of the contamination of our environment together with the widespread destruction of wildlife, giving way to what has been called 'ecological literary criticism' or 'ecocriticism'. Together with Carson, other early seminal works that have influenced the development of ecocriticism include Lynn White's essay 'The Historical Roots of Our Ecologic Crisis' (1967), which frequently serves as the starting point for literary and cultural scholars addressing environmental issues, whilst Paul Shepard's 'Ecology and Man – A Viewpoint' (1969), criticises the 'web of life' and urges that we might revive a more primitive way of perceiving humanity by redefining the self. Aldo Leopold remains to this day one of the best-known figures of applied ecology. His writing and particularly his essay 'The Land Ethic' (1949),[6] was extremely influential in conservation and the development of environmental ethics.

In Britain, Max Nicholson's *The Environmental Revolution* (1970), reads as an early account of the post-war transformation of public life and thought which Nicholson himself had partly helped to lead. However, with the exception of Jonathan Bate (2000: 27), neither American nor British ecocritics cite or consider Nicholson's book as being one of those seminal works, although few would dispute that Nicholson made some of the greatest contributions to nature conservation and the emerging environmental movement in Britain. In 1960 Nicholson chaired the committee which set up The World Wildlife Fund for Nature (WWF), the first global conservation pressure group.[7] For Hughes however, Max Nicholson was 'one of the prophets and chief publicists of this revolution' (Hughes, 1970: 19). Hughes largely overrates Nicholson's book to the point that his review was probably more noteworthy than the book itself. Hughes considers Nicholson an outstanding figure of conservation and endorses Nicholson's criticism of the government's 'mismanagement' and administration of legislation regarding natural resources, and cites, as an example of the need for the public to pressure for government intervention, 'the industrial poisoning of the water-systems in and around England' (*WP*: 131).

Hughes's review was first published in the *Spectator* (March 1970) and later collected in *Winter Pollen*, with a longer version appearing in a new magazine named *Your Environment* (Summer 1970). What has remained unrecognised is the fact that it was Hughes himself who set up *Your Environment*,[8] the first 'environmentally conscious' magazine to be published in Britain, whose main focus was on environmental issues and whose main aspiration was 'to help improve the way we all live'(Skea, 2012).[9]

It was that same compelling sense of responsibility to take things on that would later lead to his active involvement in numerous educational and environmental projects such as his wish in 1972 to set up a tree planting scheme with schools in Devon.[10] By actively engaging the children Hughes wanted to encourage a sense of empathy through learning responsibility for the respect of their immediate physical environment and instigated a plan of action by attempting to insert environmental education into public school experience.

From the 1980s onwards, Hughes's growing awareness of how nature was being degraded compelled him to take further action. Whilst in his writing his earth goddess became more clearly identified with nature, through his own activism he gave a more 'practical' form to his earth goddess worshipping. Hughes understood that environmental issues were also issues of social concern, and used his poetry and narrative to voice this, seeking through his healing quest to uncover and explore environmental repercussions of rifts between humankind and the natural world.

Whilst Hughes's early poetry illustrates the spiritual bond between the human and the non-human, it is in later poems such as 'The Black Rhino'[11] that the environmental component of Hughes's work becomes more explicit, by depicting human-induced changes to nature and the dangers of neglecting our planet. Hughes wrote the poem, which is accompanied by a long note about the decline of the species, to support an environmental campaign to help save the black rhino from extinction, the Rhino Rescue scheme in Kenya. Written out of commitment, Hughes speaks through what Vic Allen has called 'the voice of hurt' (1990: xiv). The three-part poem reads as if seen through the lens of a video camera, filmed by the reader who watches from inside the comfort of his jeep, whilst reminded that 'By the time he reaches you nothing will remain, maybe, but the horn – an ornament for a lady's lap' (Hughes, 1989: 55). The central horror of the poem is found in the simple statement 'You are the crime'. The final line of the doomed black rhino warns us that it is already too late and grieves: 'The Black Rhino/is vanishing/Into a soft/Human laugh' (Hughes, 1989: 55). In the poem Hughes portrays a species victimised by humans to near extinction through a bleak vision of death. Hughes's voice of hurt is a voice of grief, one that expresses the pain he felt over the destruction of the natural world and its species caused by modern man and through which he accuses human beings of ruthlessly destroying the environment. Moreover, by using totem figures such as the black rhino as flagship species, and at the same time helping to raise funds, Hughes invites the reader's consideration towards animals at risk of extinction.

Together with raising awareness towards threatened species, Hughes was also deeply concerned with the welfare of local flora and fauna and wrote many letters to national newspapers on such issues including the decline of otters, together with the exploitation of fishing areas, and helped raise awareness of concerns like river pollution. In addition, he played an active part in campaigning for a hygienic sewage disposal in the south-west of England by writing a reasoned campaign statement for a public enquiry and helped to found a pressure group that expanded into a national research and monitoring organisation concerned with the water quality in the nation's rivers (Gifford, 2008).

Hughes's political discourse underlying poems such as 'The Black Rhino' would now be recognised by ecocritics as environmental justice – the concern for both environment and human's dependency upon it – and can also be found in his children's poetry. The fact that the rivers of England were becoming so polluted that they contained few fish was something that Hughes had already written about in *Season Songs* (1976) in poems such as 'The River in March' where the river 'has lost all her fish. And she shivers' (*CP*: 197) or 'December River' where 'This vein from the sky is the sea-spirit's pathway', and one in which the poet finds 'only endlessly empty water' and 'backwater mills rubbish' (*CP*: 241). Already then, back in the mid-1970s, Hughes's belief that the river was a 'vein' of the earth reflected our ecological crisis, alerting us not only to the dangers of manipulating natural resources and destroying the environment, but in his depiction of the river as an indicator of human water pollution.

It is, however, in his children's writing where Hughes takes his enactment of ecological awareness a step further. In the opening pages of *The Iron Woman* (1993) the young girl, Lucy, is the first to realise that something is wrong with the fish in the river. In his children's fable the poet's frustration with governmental and political institutions who failed to respond to these issues also becomes apparent and is clearly voiced when he directs the Iron Woman's revenge on the waste disposal factory responsible for poisoning the rivers and polluting the seas. It is here where his healing quest is most clearly enacted, given that Hughes not only chooses a child as the first to identify and empathise with the problem but it is the children who understand the urgency and are able to bring about a resolution.

Parallel to his environmental campaigning during the 1980s, and his concern for river pollution, in 1984, a year after his celebrated collection, *River*, Hughes became Poet Laureate. As Gifford points out 'Hughes used the Laureateship to challenge the establishment on behalf of nature' (2011: 152). Hughes was, of course, not only aware of the effects of human pollution and the widespread use of pesticides on the environment but realised the urgency in getting this message across to the rest of society.

As Poet Laureate, Hughes wrote two election poems 'First Things First (An Election Duet Performed in the Womb by Foetal Twins)' (1987), and 'Lobby from Under the Carpet' (1992), both of which are clear examples of how by incorporating scientific data into his poetry, Hughes continued his commitment to campaigning and his ability to raise awkward questions, which in turn fed into his poetry, in the form of polemic and apocalyptic subject matters. In the first poem, commissioned by *The Times* (5 June 1987), the poet laments the predominant lack of common sense in today's society and condemns environmental issues ranging from water pollution, through crib deaths together with the atomic bomb, where the twins trust that one day, man will have enough sense to put the world to rights. The second poem also appeared in *The Times* Polling Day issue (9 April 1992), after John Major won leadership of the British Conservative Party from Margaret Thatcher. In it Hughes reiterates, with his usual dosage of bitter irony, his view that the 'Premier' should heed the results of scientific research in which he exposes toxic chemicals as the direct source of the decline in human reproductive power over the last 25 years.[12] Although Hughes's provocative poems are by no means his best, they served their didactic purpose to open debate on important themes, although they undermined to a certain extent his own political purpose: the poet's sense of responsibility as Poet Laureate, and his social concern to voice his radicalism and promote environmental awareness.

One of Hughes's most outstanding environmental essays is the introduction he wrote to *Your World* (1992), the profits of which went to the United Nations Environment Programme, which later was published as a review article in the *Observer*.[13] The poet uses his analysis here largely as a platform to 'challenge the establishment', by voicing his alarm at environmental issues such as radioactive dumping, nuclear waste, the decline of local species and river pollution. Drawing on the alienation between humankind and nature, and acknowledging the fact that we are part of something much bigger and sacred.

Hughes puts forth the same argument as David Abram's *The Spell of the Sensuous* (1996), by claiming that humans have lost all ability to understand the Earth's needs: 'Unluckily for the Earth, mankind has difficulty finding a language in which to understand her case … because modern civilisation has been able to develop only by becoming deaf to those needs' (Hughes, 1992: 30). However, according to Hughes, not only is modern civilisation to blame, since such a deafness is 'built into our governments, our industries, into all our institutions, into our brilliant culture and cleverness, into our whole more or less profitable, secure, gratifying way of life' (1992: 30).

It was of course that same detachment that Hughes attempted to heal through his holistic approach to ecology. In his children's writing his environmental responsibility and 'ecological sensibility' celebrate imagination as an important factor to change human's way of thinking. Hughes's political and intellectual ideologies were closely linked to his own inner world, requiring him to speak not only as a voice for nature, but *through* nature itself. Drawing on Carson's 'sense of wonder', based on instilling in the young child an appreciation of the natural world, together with Lawrence Buell's (1995) seminal term of 'environmental imagination',[14] referring to the reader's ability to experience a sense of connection with the environment and a deeper understanding and appreciation of the natural world, it is this separation that Hughes attempts to heal in his children's writing, by advocating a sense of connection between humans and nature.

Farms and *Moortown*

Hughes's own convictions together with his ecological concerns not only underlined the thinking behind his poetic voice but also led to his active involvement in helping charities and advocating for civic and educational projects throughout schools in Britain. One of such projects was 'Farms for City Children', a farm for children from inner cities, set up by Clare and Michael Morpurgo in 1976, in order to offer them a hands-on experience and enrich their lives.[15]

The programme involved 20 to 30 children spending a week living on Nethercott farm, in Iddesleigh, Devon, interacting with the animals and experiencing what it was like to be part of a real farm. As chance would have it, Ted Hughes was a nearby neighbour and often fished on the River Torridge, which bordered the farm nearby. Morpurgo recalls how he would come and read to the children at the farm who would sit mesmerised by his voice:

> I shan't forget his stillness as the children sat listening to him around the log fire, agog not at his reputation – they cared nothing for that and he liked that – but at the power and the fun in his words, and the music in his voice.[16]

Hughes was to become an important figure in Morpurgo's life, not only because of the genuine sympathy he showed towards the project, but also as an important source of inspiration to Morpurgo as a writer who would eventually become Children's Laureate himself from 2003 to 2005, a post that Hughes established as Poet Laureate.

However, perhaps one of the most important direct experiences related to Hughes's 'greening' was that of acquiring Moortown Farm, together with Carol Hughes, his second wife, in 1972. Some of Hughes's best-known poems, as well as some of the most graphic and disturbing come from his hands-on experiences, appear in *Moortown* (1979),[17] dedicated to Jack Orchard, whose death in 1976 brings the collection to a close. Whilst *Season Songs* (1976) manifests itself as a lyrical book of poetry primarily for young readers, both volumes are the direct product of his experiences and, in the case of *Moortown Diary* (1989), his responsibilities, as a farmer. In the Introduction, Hughes explains how the poems, written between 1973 and 1976, are simply diary entries which he returns to in order to write a poem. Written as casual notes they recorded the events of his farming days. Here he comments on how he finds himself in the position of a translator upon looking at the entry for 'February 17th':

> whatever I might make of this passage, I was going to have to destroy the original. And what was original here was not some stranger's poem but the video and surviving voice-track of one of my own days, a moment of my life that I did not want to lose.... Altering any word felt like retouching an old home movie with new bits of fake-original voice and fake-original actions.
> (*MD*: xi)

'February 17th' is the poem that stands out the most in the book. It has what Gifford has called an 'anti-pastoral' force (2009: 52) and challenges the reader against the common cuddly associations that lambs tend to have. The poem narrates in graphic detail and first person how 'A lamb could not get born'. The atmosphere of 'ice wind' already physically warns us of the tragic end that is to prevail on that pitiless February morning from the very first verse and tells how a dead lamb cannot be forced out by the mother giving birth, despite the poet's efforts. In order to save the life of the mother the poet tells in explicit detail how he has to rope the lamb's head and walk two miles to get a razor to cut off her skull, and then 'Sliced the lamb's throat-strings, levered with a knife/Between the vertebrae and brought the head off/To stare at its mother' (*CP*: 519).

The poem ends with the living mother facing her dead offspring in the mud 'the body lay born, beside the hacked-off head' (*CP*: 519) and so the cycles of life and death are brought together, as indeed they have been in the trial undertaken by the poet himself. The poem, and to a certain extent *Moortown Diary*, mark a turning point in Hughes's perception of nature and culture, in his attempt to restore human's alienated environmental consciousness through an awareness of nature. As Roberts has correctly pointed out the poem 'illustrates a new relationship between the human and animal worlds' (2006: 124), a relationship which is also present in *Season Songs* (1976) as well as in other of his children's books. In words of Roberts 'The human has a greater awareness, and may occasionally be able to help, but mostly can only look on in baffled sympathy' (2006: 124). By conceiving that 'the human has a greater awareness', Roberts is implying an anthropocentric viewpoint. Furthermore, to state that

Hughes 'mostly can only look on in baffled sympathy' in his unflinching detailed and graphic description of the dead lamb does not perhaps accurately portray the poet's responsibility. It should also be remembered that this is a practical rather than mythical test for the poet, and requires the urgency and need to take action in order to save a life. Despite exploring the connection between man and nature, this disturbing poem shatters the cosy conceptions of nature that the accommodated urban reader may have and is surely more about the disruption of comfort that it causes the reader, who is left in 'baffled sympathy' feeling physically bruised and numb inside, rather than the poet.

Critics have often overlooked the importance of the preface to *Moortown Diary*, which Hughes uses as a platform to voice his political denunciation and sense of ecological responsibility. In it the poet expresses his criticism towards the use of chemicals and technological revolutions and political pressures in the 1970s, which he considers as 'one of the biggest extinctions so far in the evolution of English countryside and farming tradition' (*MD*: ix). Like Carson's criticism of chemical farming, Hughes claims that the 'whether we liked it or not, we were in the front line of the first campaign of what felt very much like the Third World War conducted by other means – the EEC Agricultural Policy War' (*MD*: ix). He heavily criticises these policies as being the responsible party for subsidising the use of chemicals which were not only pumped into the cattle, but following the chain of life, became part of the land and were in turn passed on to the rivers and inevitably absorbed by human beings having devastating effects. As Hughes was well aware chemical farming and industrialised agriculture have a major impact on the environment. Indeed the use of chemicals in livestock and the contamination of ground water by pesticides continue today to pose a serious threat to wildlife and humans.[18]

Hughes was of course an environmental writer ahead of his time. He was deeply aware of the consciousness of environmental problems and the social change that was beginning to take place in the 1960s and 1970s, as reflected by his familiarity with pioneering works by authors such as Carson or Nicholson, alerting people to the danger of manipulating with natural resources and destroying the environment. However, even more importantly, he was already aware that the global crisis that we were facing in his time, and the environmental problems like those Carson had referred to, were largely of our own making. Attempting to give the answer to why we are in a state of crisis with the global environment Donald Worster in *The Wealth of Nature* explains that we are immersed in a global crisis today 'not because of how ecosystems function but rather because of how our ethical systems function' (1994: 27). As he explains:

> Getting through the crisis requires understanding our impact on nature as precisely as possible, but even more, it requires understanding those ethical systems and using that understanding to reform them. Historians, along with literary scholars, anthropologists, and philosophers, cannot do the reforming, of course, but they can help with the understanding.
>
> (Worster, 1994: 27)

Worster's ideas appear to echo those of Hughes, with the exception that Hughes would, of course, add writers for children to this list.

Hughes's approach to ecology and the goddess of nature

The therapeutic function of healing that Hughes ascribes to nature is closely linked to the development of Hughes's sense of environmental responsibility. Hughes's approach to ecology is at large a holistic one which expresses similar concerns to those of what might be called 'deep ecology', 'social ecology' and ecopsychology. His environmental responsibility and 'ecological sensibility' celebrate imagination as an important factor to change human's way of thinking. The underlying ecological approach that informs much of Hughes's children's writing, the idea that environmental issues are also social issues, is close to what ecocritics now call 'social ecology'.

Despite similar concerns, ecologists have long been divided in how they view human's place and role in relation to nature. One of the first to focus on this distinction was Murray Bookchin, the originator of what he called 'social ecology'. Bookchin's central argument is that the destruction of nature originated in hierarchical and class-structured social domination amongst humans and is based on the conviction that the environmental crisis is a result of the Western ideology of dominating the natural world (Bookchin, 1982). The alternative is a society based on ecological principles and mutual respect, underlying the principle that if we change human society then our relationship with nature will also become transformed. In Bookchin's view, social ecology is not only a powerful discipline from which to draw a critique of the present social and ecological disorder, but more importantly it 'provides more than a critique of the split between humanity and nature; it also poses the need to heal them. Indeed it poses the need to transcend them' (1982: 22). Following Bookchin, John Clark describes social ecology as 'the awakening earth community reflecting on itself, uncovering its history, exploring its present predicament, and contemplating its future' (1998: 416). Like Bookchin, Clark also emphasises understanding the relationship between nature and society from a holistic approach. Ecological problems cannot be fully understood, let alone resolved, without facing the social issues that lie behind them.

In similar terms, the renowned Norwegian eco-philosopher and activist Arne Naess, founder of the deep ecology movement, made the same distinction in a 1973 essay contrasting the 'shallow' reform environmental movement with the emerging 'deep, long-range ecology movement'. Adherents of the deep ecology movement oppose technological and managerial solutions since these constitute another form of the human-centred value system that lies at the core of Western industrial culture. Deep ecologists argue that environmental philosophy must recognise the values that are inherent in nature independently of human needs and desires, referring to a more holistic environmental philosophy. They advocate instead a biocentric view, and argue that environmental philosophy must recognise the values that lie inherent in the non-human world, independently of human wants or needs. However, radical critics of deep ecology, like Timothy Luke, call for a more detailed form of ethical thinking and consider that ecocritics can in fact *manipulate* our conception of the environment. Luke reconsiders deep ecology as a political philosophy whilst arguing that it 'has no program. At best it only offers the traditional solution of social anarchism, changing the self to change society with individual acts of will guided by correct conscience/consciousness in a pressing situation of necessity to save humanity' (1997: 24).

Some deep ecologists consider that there are close connections between certain streams in world religions and deep ecology. David Abram describes deep ecology as a way of thinking that distinguishes itself from conventional environmentalism as 'Rather than applying various "band-aid" solutions to environmental problems, adherents of deep ecology ostensibly ask "deeper" questions, and aim at deeper, more long-range solutions' (2005: 469). Abram prefers to use the term 'depth ecology', to acknowledge the fact that we neglect the depth and dimension of the world, and that we are part of something much bigger and more sacred.

Bearing in mind that the whole of Hughes's work is largely inspired by the myth of Graves's *The White Goddess* (1948), and that Hughes's goddess is above all feminine, an ecofeminist reading of Hughes's children's writing can offer insight into the way in which Hughes represents nature as well as the close relation Hughes depicts between his female characters and nature.

Ecofeminist activism grew during the 1980s and 1990s amongst women from anti-nuclear, environmental and feminist movements. Though the term 'écoféminisme' was first used by Françoise d'Eaubonne, in *Le féminisme ou la mort* (1974), ecofeminism is used as a term to name a growing activist and academic, cultural and intellectual movement which has tried to connect theory with practice, in order to come to terms with today's environmental disaster. In this sense ecofeminism is the name given to a variety of positions that have roots in the union of the radical ecology movement, 'deep ecology' and feminism. These different perspectives reflect not only different feminist perspectives but also different ways of understanding nature as well as offering solutions to our environmental crisis. As Bate has correctly stated, 'In contrast to those feminists who seek to denaturalize traditional images of masculinity and femininity, ecofeminists reappropriate and celebrate the idea of woman's closeness to the rhythms of mother earth' (2000: 107). However, by claiming that women are better nurturers and closer to nature than men, Bate also runs the risk of being criticised for falsely essentialising women. Charges of essentialism, usually used by mainstream feminists, have been made in the past towards ecofeminism for reproducing a feminisation of nature through terms such as 'Mother Earth' or 'Mother Nature'. Such terms are deemed problematic in the sense that, by gendering nature as a mother or a goddess, we are endowing females with a special relationship to nature. As Victoria Davion in 'Is Ecofeminism Feminist?' has pointed out, 'a truly feminist perspective cannot embrace either the feminine or the masculine uncritically, [but] requires a critique of gender roles, and this critique must include masculinity and feminity' (1994: 9). However, despite the existence of a problematical divide between theoretical ecofeminism and activist ecofeminism, there is, of course, the danger that any attempt to interweave feminism and ecology is inherently essentialist. Garrard suggests that radical ecofeminism's 'irrationalism and essentialism are serious limitations' to its use as a critical philosophy (2011: 30).

Despite the fact that Hughes celebrates the representation of the female through the form of his earth goddess, the use of terms such as 'Mother Nature' and 'The Goddess of Complete Being' do not necessarily imply or embrace a simplistic or essentialist notion; it would appear that Hughes risks essentialism in order to defeat Western ideas of patriarchy as will be discussed later. This also means taking into consideration the environmental values that he portrays in his work, as well as looking at the way nature is represented and the interconnections between nature and culture, in order to discuss to what extent Hughes successfully achieves a non-dualistic space for rethinking culture–nature relations.

Both social ecology and ecofeminism have criticised deep ecology for blaming humanity for the ecological crisis – especially ordinary 'consumers' and 'breeders of children' – reducing human beings to a simple species that is 'overpopulating' our planet and 'devouring' its resources whilst largely ignoring the roles played by social and economic factors inherent in global capitalism (Bookchin, 1988: 13). Bookchin strongly challenged the political perspective of deep ecology and argued that this view prevents them from seeing the real *cultural* causes of environmental problems (1988: 18). Whilst social ecologists maintain that deep ecology overemphasises cultural factors, ecofeminists have criticised the oppression of nature and women by patriarchal power structures, claiming that deep ecology fails to consider how patriarchal attitudes and institutions help to generate environmental problems.

The healing function of nature was of course central to Hughes's work for children. This holistic sense of literature is closely linked to his evocation of the goddess of nature. Hughes's own spiritual goddess of nature draws on the ancient idea of the mother goddess whose body constituted the Earth and shares with ecofeminism similar concerns with the relationship of scientific worldviews to religion. Most ecofeminists would argue that Western science has projected a material idealised world that is not sacred, but mechanistic. There are also those who have positioned themselves as spiritual ecofeminists, such as Starhawk, embracing the religious earth goddess and defining her theistic spirituality in opposition to Judeo-Christian theism. Starhawk (1979) maintains that there is a sacred dimension immanent in nature and argues that the separation of the divine from nature is one of the major causes of the ecological crisis.

Taking an ecofeminist approach to Hughes's concept of nature, Hughes's way of thinking about human–animal relations holds promise for healing the bonds between humans and nature, and draws on the roots of ecopsychology and David Abram's inspiring book *The Spell of the Sensuous* (1996). Simply put, ecopsychology is situated at the crossroads of environmental philosophy, psychology and ecology, and explores the human–nature relationship.[19] Like Hughes and Carson, Abram suggests that there is a synergistic relation between the human and non-human; one which begins in childhood but usually fades with adulthood, through our disengagement and alienation with nature. Abram's book can also help to inform an understanding of Hughes's dichotomies of culture and nature since Abram shares related concerns to those of Hughes and raises similar questions:

> How did western civilization become so estranged from nonhuman nature, so oblivious to the presence of other animals and the earth, that our current lifestyles and activities contribute daily to the destruction of whole ecosystems – whole forests, river valleys, oceans – and to the extinction of countless species?
>
> (1996: 137)

According to Abram it is by our disengagement with the world, and with it the process of alienation that we have become divided from the realms of our own animal perceptions and separated from the reality of our senses. In order to heal our alienation, Abram's message – which can also be read in much of Hughes's children's writing – is to re-immerse ourselves in nature and awaken our senses, by putting our inner world in contact with the outer world.

Notes

1 *Silent Spring* was first published in the USA in 1962 and the British version was first published in 1963.
2 The two books Hughes is referring to are Carson's *Under the Sea-Wind* (1941) and *The Sea Around Us* (1951).
3 The book was first published as an article 'Help Your Child to Wonder' in *Woman's Home Companion* in July 1956 and later published posthumously by Harper & Row as *The Sense of Wonder*, in 1965.
4 The original interview by Nigel Forde was first broadcasted in *Bookshelf*, BBC Radio 4, 20 March 1992.
5 Ted Hughes wrote a footnote to Nigel Forde's interview in a letter to Terry Gifford on 17 December 1993.
6 'The Land Ethic' was written in 1947 and first published in Leopold (1949).
7 According to the Deed of Foundation, the organisation was created to ensure the

> conservation of world fauna, flora, forests, landscape, water, soils and other natural resources by the acquisition and management of land, research and investigation, education at all levels, information and publicity, coordination of efforts, cooperation with other interested parties and all other appropriate means.
>
> (Callicott and Frodeman, 2009: 412)

8 *Your Environment* was launched in 1969 by Hughes, together with David Ross and Daniel Weissbort and ran for over three years.
9 The magazine was launched in 1969 by Hughes, together with David Ross and Daniel Weissbort and ran for over three years. Ross became editor and was joined the following year by a young Canadian scientist, Walter C. Patterson, who took over Ross's role as editor whilst Weissbort and Hughes remained advising and contributing

editors. Ann Skea (2008) cites that amongst the topics that constituted the subject matter of the magazine 'were household issues such as non-returnable bottles, toxic detergents, packaging, and garden chemicals; a 'Whitehall Diary' recording environmental issues discussed in parliament; the water supply; defoliants; transport; and science news' (n.p.). Skea also quotes Daniel Weissbort who remembers that the magazine was 'a money loser', but that it was 'very much part of the ethos of the time' (2008: n.p.).

10 See Hughes's article in the *Times Educational Supplement* 'Children Plant Trees for Tomorrow', 25 October 1972. Further documentation referring to this project can be found in correspondence between Hughes and the *Times Educational Supplement*, and letters from Trees for People and the Devon Tree Bank.

11 Published in *Wolfwatching* (London: Faber and Faber, 1989). This poem was written for HRH The Duke of Edinburgh as President of the World Wildlife Fund (now known as the World Wide Fund for Nature), as a contribution to the Rhino Rescue Appeal of 1987.

12 Hughes largely refers here to John Elkington's book *The Poisoned Womb* (1985) and to scientific results on the testicular histology and dysgenesis (sterility) referring to the effects of environmental pollutants on the human reproductive cycle, showing the fall in sperm count.

13 Hughes reviewed *Your World* for the *Observer Magazine*. To coincide with the Earth Summit and its own twentieth anniversary, the United Nations Environment Programme (UNEP) together with Canon Inc. held a photographic competition based on the theme 'Focus on Your World'. More than 30,000 photographs were submitted from more than 140 countries resulting in the largest international photographic competition ever to be staged at that time. The book contains photos on environmental themes ranging from categories crossing animals, seas, environment, nature and people all focusing on the vision participants had of the world either in celebration or condemnation of its present condition. The review article was published on 29 November 1992.

14 Buell's term comes from his book *The Environmental Imagination: Thoreau, Nature Writing and the Formation of American Culture*, published by Harvard University Press (1995).

15 Like Hughes, Michael Morpurgo before becoming a writer, worked as a teacher in the 1960s, and felt disillusioned and frustrated with the education system. As Morpurgo himself explains:

> Clare and I had felt that many children we were teaching suffered from a profound poverty of experience, that school could so often be narrow and restricting for them, that their horizons needed expanding, that their lives needed to be enriched … time spent in the countryside, away from school and family, could only be beneficial for them.
>
> (2010: 6)

16 From 'A Celebration', Sound Archive C927/381, The British Library, London. Recorded 2008-10-20.

17 Hughes first published his collection of poems within the larger collection of several sections as *Moortown* in 1979, but the title sequence of the collection was published separately ten years later as *Moortown Diary* (1989) with an added preface and extensive endnotes to the poems. All quotes here are from Hughes's *Collected Poems* (2003).

18 Other negative effects include the loss of soil fertility and the endangerment of wildlife. According to the National Research Council (NRC, 1989), 'Agriculture is the largest source of nonpoint water pollution' (in Callicott and Frodeman, 2009: 31). Since the development of industrialised agriculture and farming in the twentieth century we have witnessed a transformation from traditional farming techniques to highly mechanised and intensive farming. Furthermore, as is well documented,

factory farming uses huge amounts of natural resources, generating tonnes of waste that contaminate and pollute the environment contributing to climate change.

19 The term was first coined by Theodore Roszak in *The Voice of the Earth* (1992). Since then it has developed as a field in itself. For a broader overview see The International Community for Ecopsychology (www.ecopsychology.org/) or The European Ecopsychology Society (www.ecopsychology.net/) or find related articles on the online *Ecopsychology Journal* (www.liebertonline.com/eco).

Archive material

'A Celebration', Sound Archive C927/381, The British Library, London. Recorded 2008-10-20.

References

Abram, David, *The Spell of the Sensuous: Perception and Language in a More-than-Human World* (New York: Vintage Books, 1996).

Abram, David, 'Depth Ecology', in Bron Taylor (ed.), *The Encyclopedia of Religion and Nature* (London and New York: Continuum International, 2005), pp. 1023–1026.

Allen, Vic, 'Truth and the Painful Voice of Hurt', *Sunday Telegraph* (18 November 1990), p. xiv.

Bate, Jonathan, *The Song of the Earth* (London: Picador, 2000).

Bookchin, Murray, *The Ecology of Freedom: The Emergence and Dissolution of Hierarchy* (Palo Alto, CA: Cheshire Books, 1982).

Bookchin, Murray, 'Social Ecology versus Deep Ecology', *Socialist Review*, 88:3 (1988), pp. 11–29.

Buell, Lawrence, *The Environmental Imagination: Thoreau, Nature Writing and the Formation of American Culture* (Cambridge: Harvard University Press, 1995).

Callicott, J. Baird and Robert Frodeman (eds), *Encyclopedia of Environmental Ethics and Philosophy* (New York: Macmillan, 2009).

Carson, Rachel, *The Sea Around Us* (New York: Oxford University Press, 1951; new edition, Oxford: Oxford University Press, 1991).

Carson, Rachel, *Silent Spring* (Boston, MA: Houghton Mifflin, 1962; London: Penguin, 1999).

Carson, Rachel, *The Sense of Wonder* (New York: Harper & Row, 1965; New York: HarperCollins, 1998).

Carson, Rachel, *Under the Sea-Wind* (New York: Oxford University Press, 1941; centennial edition, New York: Penguin, 2007).

Clark, John, 'A Social Ecology', in M. E. Zimmerman, J. Baird Callicott, John Clark, Karen J. Warren and George Sessions (eds), *Environmental Philosophy: From Animal Rights to Radical Ecology* (Upper Saddle River, NJ: Prentice Hall, 1998), pp. 416–440.

Davion, Victoria, 'Is Ecofeminism Feminist?', in Karen J. Warren (ed.), *Ecological Feminism* (New York: Routledge, 1994), pp. 8–28.

Garrard, Greg, *Ecocriticism* (London: Routledge, 2011).

Gifford, Terry (ed.), *Green Voices: Understanding Contemporary Nature Poetry* (Manchester: Manchester University Press, 1995).

Gifford, Terry (ed.), 'Rivers and Water Quality in the Work of Brian Clarke and Ted Hughes', *Concentric*, 34:1 (2008), pp. 75–91.

Gifford, Terry (ed.), *Ted Hughes* (London: Routledge, 2009).

Gifford, Terry (ed.), *The Cambridge Companion to Ted Hughes* (Cambridge: Cambridge University Press, 2011).
Graves, Robert, *The White Goddess: A Historical Grammar of Poetic Myth* (London and Boston: Faber & Faber, 1948); New York: Farrar, Straus & Giroux, 1966).
Hughes, Ted, *The Iron Man* (London: Faber & Faber, 1968).
Hughes, Ted, 'The Environmental Revolution', review of Max Nicholson, *The Environmental Revolution* (London: Hodder & Stoughton, 1970), *The Spectator* (21 March 1970), p. 19.
Hughes, Ted, 'Children Plant Trees for Tomorrow', *Times Educational Supplement* (25 October 1972).
Hughes, Ted, *Season Songs* (London: Faber & Faber, 1976).
Hughes, Ted, *Moortown* (London: Faber & Faber, 1979).
Hughes, Ted, *River* (London: Faber & Faber, 1983).
Hughes, Ted, 'First Things First (An Election Duet Performed in the Womb by Foetal Twins)', *The Times* (5 June 1987).
Hughes, Ted, *Moortown Diary* (London: Faber & Faber, 1989).
Hughes, Ted, *Wolfwatching* (London: Faber & Faber, 1989).
Hughes, Ted, 'Introduction', in *Your World* (London: HarperCollins, 1992). Also published in the *Observer Magazine* (29 November 1992), pp. 30–39.
Hughes, Ted, 'Lobby from Under the Carpet', *The Times* (Polling Day issue, 9 April 1992).
Hughes, Ted, *The Iron Woman* (London: Faber & Faber, 1993).
Hughes, Ted, *Winter Pollen* (London: Faber & Faber, 1994).
Hughes, Ted, *Collected Poems* (London: Faber & Faber, 2003).
Kerslake, Lorraine, 'Beyond Rachel Carson's "Sense of Wonder": Advocating Environmental Awareness in Ted Hughes's Writing for Children', *The Ted Hughes Society Journal*, V:I (2016), pp. 65–73.
Lear, Linda, '50 Years of *Silent Spring*', 2012. Web. 25 April 2012. http://sagharboronline.com/sagharborexpress/arts/50-years-of-silent-spring-17251.
Leopold, Aldo, *A Sand County Almanac and Sketches Here and There* (New York: Oxford University Press, 1949).
Luke, Timothy, *Ecocritique: Contesting the Politics of Nature, Economy and Culture* (London: University of Minneapolis Press, 1997).
Morpurgo, Michael, 'Once Upon a Life', the *Observer Magazine*, 11 July 2010. Web. 15 March 2013. www.theguardian.com/lifeandstyle/2010/jul/11/once-upon-life-michael-morpurgo-farms-city-children.
Naess, Arne, 'The Shallow and the Deep, Long-Range Ecology Movement: A Summary', *Inquiry*, 16 (1973), pp. 95–100.
Plath, Sylvia, *Letters Home: Correspondence 1950–1963*, edited by Aurelia Plath (London: Faber & Faber, 1976).
Reid, Christopher (ed.), *Letters of Ted Hughes* (London: Faber & Faber, 2007).
Roberts, Neil, *Ted Hughes: A Literary Life* (Basingstoke: Palgrave Macmillan, 2006).
Shepard, Paul, 'Introduction: Ecology and Man –Viewpoint' in Paul Shepard and Daniel McKinley (eds), *The Subversive Science* (Boston: Houghton Mifflin Co., 1969).
Skea, Ann, '"Notes and Queries" Related to "Environment and Society"', 2008. Web. 14 January 2012. http://ann.skea.com/NQ5eco.htm.
Starhawk, *The Spiral Dance* (San Francisco: Harper, 1979; 2nd edition, 1989).
White, Lynn, Jr., 'The Historical Roots of Our Ecologic Crisis', *Science*, 155 (10 March 1967), pp. 1203–1207.
Worster, Donald, *The Wealth of Nature: Environmental History and the Ecological Imagination* (New York: Oxford University Press, 1994).

Part II

Correcting culture's error

An ecocritical analysis of Hughes's writing for children

3 The seeds of Hughes's children's writing

An overview of the existing critical studies of Ted Hughes's writing for children

There is surprisingly little critical writing on Hughes's children's literature, both in the field of Hughes's scholarship as well as in the field of children's literary criticism. One of the first critics to address Hughes's children's writing was Keith Cushman who claimed that the poet had 'demonstrated a sustained commitment to children's literature from the early stages of his career' (1983: 240). Cushman, however, falls short of demonstrating how and why. Likewise, despite his insightful claim that Hughes thought that 'if there is a real hope, it is with the children' (1983: 240), he fails to reveal why the poet believed this. Cushman's entire chapter is in fact much of an appraisal of what Keith Sagar had already stated about Hughes's children's writing in *Ted Hughes* (1972).[1] In his reproach Sagar feels he should excuse Hughes's writing by saying that at this early stage of his career 'the poet's shaping imagination is not fully engaged' (1972: 15) and condemns *Meet My Folks* (1961) as being 'fresh and engaging, but shapeless' (1972: 15) and *The Earth Owl and Other Moon People* (1963) as 'slapdash – mere poetic doodles' (1972: 15–16). Despite both Cushman and Sagar's bleak reviews of Hughes's first books of children's poetry, both *Under the North Star* (1981) and *Season Songs* (1976) come off unscarred. Possibly because, in Cushman's words 'the most interesting question about Hughes' children's poetry is the relationship between this poetry and his poetry for adults' (1983: 241). According to Cushman 'The themes and the visionary intention are absolutely the same, and often so is the subject matter' (1983: 254), although he fails, again, to tell us what those themes are or what the subject matter may be. Cushman then bluntly concludes by quoting Hughes himself and declaring that Hughes's children's poems are written 'within hearing of children' (1983: 255), at the same time as he underlines that according to Hughes 'children are capable of comprehending and responding to adult poetry' (1983: 255). That Hughes wrote 'within hearing of children' is a crucial phrase and one which will be discussed later. Perhaps, however, the question Cushman should have been raising here was whether or not Hughes thought that adults were capable of responding to children's poetry and not the other way round.

Following Cushman's early critical writing on Hughes's works for children, the most prominent voices and major contributors in Hughes's critical studies that have successfully focused on Hughes as a children's writer belong to leading children's literature critic Lissa Paul and independent scholar Claas Kazzer.

Lissa Paul's work explores Hughes's children's writing largely as a continuum of his writing for adults.[2] Paul, unlike Cushman, claims that not only does Hughes write for children but that he does so in a 'lingua franca', one that was suitable for secretly listening adults, and not, as Cushman argues, the other way around. Paul was also one of the first to embrace the healing aspect of Hughes's writing by considering *The Iron Man* 'a redemptive story' (2000: 225) and linking Hughes's need at the time of writing the story to his personal suffering and that of his children after Sylvia Plath's suicide. Not only did Hughes correspond with Paul, but he answered her questions concerning his writing for children and adults, and presented her with some of his most original and personal statements and insight into his children's writing.[3]

Claas Kazzer, an independent scholar, runs a web site for scholars on Ted Hughes (see www.ted-hughes.info) containing news and information about articles and publications, a brief biographical sketch together with many links, and has also written about his books for children. In 'What is the Truth? Ted Hughes, Childhood, Memories and Stories' (2010), Kazzer looks at the way in which Ted Hughes's childhood experiences and places are reflected in his poems and stories, and raises the question of Hughes's conflicting narratives as well as analysing the inconsistencies that appear in many of his stories. In his article Kazzer underlines the transitions between biography and fiction in his tales, especially in the case of *Difficulties of a Bridegroom* (1995) and shows how several of the stories, including 'The Deadfall' are based on autobiographical events that took place around Mytholmroyd. Like Paul, Kazzer also considers that Hughes's writing for children is connected to his adult writing, underlining that at the heart of Hughes's children writing, as in his writing for adults, lies Hughes's myth of the Goddess (of Complete Being), whose laws shape and govern Hughes's characters, an aspect which other critics who have researched his children's writing have failed to acknowledge.[4] Moreover Kazzer correctly considers that Hughes's writing for children presents us with a more open and personal view of the poet, as well as offering Hughes a way to break free from his adult concerns and that in it Hughes was 'giving away some intimate part of himself'. Kazzer believes that 'at the most personal, creative level of this writing, we encounter a poet with a near obsession with healing, redemption, with the balancing of "male" and "female" aspects of one's being, with making good and resurrecting' (Kazzer, 2000).[5] This redemptive power of writing and the idea of 'balance' in his children's writing are points that will be later discussed at length.

Following early critical writing on Hughes's works for children other, more recent studies include Neil Roberts's chapter 'Writing for Children' in *Ted Hughes: A Literary Life* (2006) which gives an excellent overview of his children's writing. He begins by reminding us that 'Hughes wrote more books for children than collections of original poetry for adults' (Roberts, 2006: 167) and distinguishes three phases in his work through which he looks at the themes and styles of his major works, giving special attention to *The Iron Man* (1968) and its sequel *The Iron Woman* (1993) which Roberts describes as 'a much more urgent, and didactic, ecological intervention' (2006: 177).

Other existing critical writing on his children's work include a handful of articles published in academic children's literary journals or newspapers, and several book chapters following that of Cushman's. These include Blake Morrison's 'Man of Mettle' (1993), Anthony Wilson's 'Ted Hughes's Poetry for Children' (2001) and Neil Philip's insightful article about *What is the Truth?* in 'The Shawl of the Beauty of the World: The Children's Books of Ted Hughes' (2003). More recent publications include Michael Lockwood's 'Ted Hughes: The Development of a Children's Poet' (2009), which looks at how Hughes's poetry for children developed over the years and correctly argues that his later poems represent an undervalued phase of Hughes's work which have rarely been discussed by critics. Children's poet and writer Mick Gowar has recently written about the symbiotic relation between Hughes and Leonard Baskin and used Hughes's writing for creative workshops for children and adults.[6] Peter Cook's chapter 'Poets in the Making: Ted Hughes, Poetry and Children' in *Poetry and Childhood* (2010) considers *Poetry in the Making* (1967) as being a central piece to understanding Hughes's life-long commitment to involving young people in poetry and assesses its effectiveness as a stimulus for children's understanding and enjoyment for writing poetry. Perhaps David Whitley in his essay '"The Fox is a Jolly Farmer and We Farm the Same Land": Ted Hughes and Farming' (2013) has been most successful in comparing Hughes's children's writing with his adult poetry whilst looking at his underlying engagement with the natural world linked to his work as a farmer.

Following several international conferences on Hughes's work, and thanks to the creation of the Ted Hughes Society,[7] critical attention appears more recently to have shown a new interest in Hughes's children's work, especially in relation with his writing for adults. Jonathan Bate's recent biography *Ted Hughes: The Unauthorised Life* (2015) has of course also stirred fresh interest in Hughes. However, despite its 622 pages, not only is there little room for Hughes's environmental and educational work, but his children's literature is barely acknowledged either.

Meanwhile acknowledgements to Hughes's contribution to children's literature from major children's literature scholars have remained somewhat mixed. On the one hand in Britain renowned children's literature critic Morag Styles has dedicated a chapter to Hughes's 'great poetry of childhood' in *From the Garden to the Street: Three Hundred Years of Poetry for Children* (1998). Leading children's critic Peter Hunt gives Ted Hughes a somewhat brief acknowledgement in *Children's Literature: An Illustrated History* (1995) and has dedicated an entry to the writer in *Children's Literature* (2001), acknowledging that 'his children's poems are generally (politely) ignored by adult literary critics and yet inordinately praised by children's literature critics' (2001: 89). However, on the other hand most major international children's literary critics from outside Britain, mainly from USA, fail to mention Hughes as a children's writer. Seth Lerer in *Children's Literature: A Reader's History from Aesop to Harry Potter* (2008) gives no mention of Hughes in the index. Hughes's name fails to appear on any of the 768 pages that form the colossal *Children and Books* (1991) edited by Zena Sutherland. Nor is there any mention of Hughes in Anita Silvey's much praised comprehensive survey of contemporary children's literature *The Essential Guide to Children's Books and Their Creators* (2002). Jack Zipes's *The Norton Anthology of Children's Literature* (2005) does include the first chapter of *The Iron Giant: A Story in Five Nights*,[8] although this was probably largely due to the fact that Lissa Paul co-edited it, since in none of Jack Zipes's critical studies of children's literature is there any mention of Hughes, although curiously enough Plath's name does appear in all the aforementioned books.[9] His suitability as a children's poet was even put to question in 1979, when he was awarded the *Signal* Poetry Award for *Moon-Bells*, by critics who debated whether his poetry was too difficult for children. It seems almost as if the American establishment had agreed through a sort of silenced conspiracy, in line with angry feminist attacks in the late 1970s, to discredit Hughes, and punish him by omitting his name from the list of children's writers.[10]

In summary, it would appear that a focus on the literature that Hughes wrote for children is largely missing from the existing canon of criticism of Hughes's work together with that of major Hughes critics, who have largely ignored the link between his critical writing and his imaginative writing. Perhaps due to the fact that the landscape of Hughes's early adult writing was largely filled with figures of shamans and tricksters, critics have often described his work as a shamanic narrative (Faas, 1980; Hirschberg, 1981; Sagar, 1978), or a trickster one (Bradshaw, 1979). However, other authors (Gifford and Roberts, 1981) already appeared to be aware of the struggles that created dialectical tension in Hughes's writing, and related them to sets of oppositions and binary dualities such as nature and culture or imagination and reason. All of these studies provide insight into the poet himself, but fail to explore the scope of his variety of works for children as such. Moreover, there appears to be a general lack of discussion amongst Hughes's critics about the way in which his children's writing arises in connection with the inner world of the poet.

Hughes's cauldron of ideas: from *Poetry in the Making* to 'Myth and Education'

Shortly after moving with Plath to Primrose Hill, London, on 1 April, 1960 Plath and Hughes's first child, Frieda Rebecca, was born, followed by their second, Nicholas, on 17 January, 1962. Partly to support his new family, Hughes began working with Moira Doolan, of the British Broadcasting Corporation (BBC) Schools Broadcasting Department, on a series of radio programmes of readings of verse and prose, recorded and broadcast between 1961 and 1965. The broadcasts were a wave of fresh air from the rather dull programmes offered and were eagerly anticipated by both children and teachers from primary and secondary schools alike. The scripts that Hughes wrote for the series 'Listening and Writing' were to become his seminal book *Poetry in the Making*, but not until 1967. It seems only fair that Hughes should have dedicated his book to Pauline Mayne and John Fisher, his teachers from Mexborough Grammar School, who had encouraged both his reading and creative writing. In a letter to John and Nancy Fisher, on 11 March 1967, Hughes states that his work 'falls between all stools' (*LTH*: 92), endorsing the idea that Hughes wrote 'within hearing' of both an adult and child audience. In it he sets out collecting his ideas in a kind of handbook both for adults, especially teachers, and of course with a young audience in mind.

It is also one of Hughes's few books which both Hughes scholars and children's critics have generally agreed on, and praised, despite having different reasons for doing so. Morag Styles, in *From the Garden to the Street* (1998), claims that *Poetry in the Making* 'is an excellent book about reading and writing poetry … This classic text is addressed to children, but avidly read by teachers who want to learn more about using poetry in the classroom' (253). Keith Sagar considers *Poetry in the Making* 'the best book I know about the writing (and reading) of poetry in schools' (1978: 18). Sagar however, appears to later change his mind about the intended audience when he comments that 'Most of the exercises in *Poetry in the Making* probably had their origin as exercises devised primarily for Plath' (2006: 58) and claims that 'The final image is clearly much more appropriate to Plath than to a class of schoolchildren, and reveals the true origin of these exercises' (2006: 59). Sagar is most probably referring here to Hughes's genuine belief in writing exercises, which he provided both to Plath and his sister Olwyn[11] and which he describes in detail for teachers to use in the classroom in *Poetry in the Making* (1967). Although Sagar is right in saying that Plath was *influenced* by Hughes's words, when she wrote in 1957 'All I need to do is work, break open the deep mines of experience and imagination, let the words come and speak it all, sounding themselves and tasting themselves' (in Sagar, 2006: 58),[12] it would seem that Plath is not echoing Hughes's voice here, but rather searching for her own. This is a very personal, and almost desperate, voice speaking about her feelings of inferiority with respect to her husband, who was already an acknowledged poet at the time. If anything, Hughes's writing of *Poetry in the Making* was more an attempt on Hughes's behalf to show *his* own personal voice, and in order to share with the reader both the pain and joy experienced in his work he reveals the naked truth of his own writing. By going back to his childhood roots, Hughes was storying his life, through his own experiences and memories. That is, he was drawing on what had fed his own imagination in order to articulate the guidance, and begin his own both personal and public healing quest, as a shaman, through the power of poetry.

However, the real importance of *Poetry in the Making* lies not in the question of Hughes's intended audience, but rather in the fact that it was central to Hughes's life-long commitment to involving young people in writing. It should not be forgotten that *Poetry in the Making* was also a therapeutic way for Hughes to deal with his terrible loss after Plath's suicide whilst continuing to nurture children's creative instinct and imagination. In it he shows how poetry has the power to heal both the poet and the reader. Furthermore, the fact that the text retains the oral quality of its traditional broadcast version, allows the reader to imagine that one is listening to Hughes's own powerful voice. Hughes begins by recalling not how he came to write, but how he came to capture animals, exposing his own naked childhood core:

> There are all sorts of ways of capturing animals and birds and fish. I spent most of my time, up to the age of fifteen or so, trying out many of these ways, and when my enthusiasm began to wane, as it did gradually, I started to write poems.
>
> (*PM*: 15)

The poet then goes on to explain how capturing animals and writing poems have been 'one interest' (*PM*: 15) and the connection that hunting, fishing and writing have for him. He confesses that animals 'know something special ... something perhaps which we are very curious to learn. Maybe my concern has been to capture not animals particularly, and not poems, but simply things which have a vivid life of their own' (*PM*:15). Hughes takes the reader by the hand, leading us into his own central childhood experience, to show us his own view of poetry: the fact that, for the poet, poetry is *alive*.

As David Whitley in his essay 'Animals in Children's Poetry' puts it: 'Capturing animals in verse is a project directed as much towards enabling humans to reconnect with their real animal core as it is towards garnering a renewed sense of respect and responsibility for a threatened and diminishing wild nature' (2010: 191). Whitley identifies how Hughes captures animals as a way of connecting our inner self with our 'animal core', and uses the concept of 'redemption', although he fails to link it to Hughes's own healing quest. Hughes achieves this 'redemption' in his writing by reconnecting dualities such as that of the human/non-human, or nature/culture in order to attain a sense of balance by enabling the reader to reconnect and overcome the gap between the human and non-human. In doing so the poet projects a redemptive aspect: that of allowing the reader to empathise with nature and the non-human.

In the introduction to *Poetry in the Making*, Hughes claims that an integral part of his method is 'not "How to write" but "How to try to say what you really mean" – which is part of the search for self-knowledge and perhaps, in one form or another, grace' (12). Hughes's choice of the word 'grace' is related here not only to the Romantics' idea of childhood, but it also conveys a redemptive aspect. Hughes then goes on to explain his method in a language of sensory awareness and shows the reader how to make the poem come alive:

> Imagine what you are writing about. See it and live it. Do not think it up laboriously, as if you were working out mental arithmetic. Just look at it,

> touch it, smell it, listen to it, turn yourself into it. When you do this the words look after themselves, like magic.
>
> (*PM*: 18)

To illustrate this, Hughes cites his 'first animal poem': 'The Thought Fox',[13] a poem 'about a fox, obviously enough, but a fox that is both a fox and not a fox' (*PM*: 20). That is, a poem which captures the spirit of a fox. 'The Thought Fox' indicates most clearly Hughes's poetic mission at the same time as it illustrates how he sought to 'capture' it: through the engagement of our imagination. Indeed, by offering the reader access into the inner world of spirits, and by engaging our imagination at the deepest levels of experience the poet is connecting not only the human and animal world, but also what he has called the inner and outer world. As Hughes explains in 'Myth and Education':

> The inner world, separated from the outer world, is a place of meaningless objects and machines. The faculty that makes the human being out of these two worlds is called divine.... More essentially, it is imagination which embraces both outer and inner worlds in a creative spirit.
>
> (*WP*: 151)

For the poet the 'outer' world is a world of objective reality, and in order to achieve wholeness we must learn to connect with our 'inner' world. That is, a world that is linked to our feelings, by means of imaginative patterns and myths. Hughes is not advocating a return to religion, but rather a faculty which embraces both worlds and that faculty is, of course, imagination. Since education appears to have neglected this faculty, Hughes's first concern in *Poetry in the Making* is to foster imagination by engaging the reader actively in the creative process.

Following this, in the second chapter, the poet explains the importance of memory and experience, in relation to our imaginative core, by showing how poetry is made out of experience. Throughout his autobiographic testament Hughes stories his own experience in order to tell his own truth: his own very personal myth, and so it is by storying his experience that he reaches his childhood core.

Throughout the book Hughes explores a range of themes, drawing inspiration from both the natural world, as in 'Capturing Animals', 'Wind and Weather' or 'Writing about Landscape' and the world of fantasy (as in his chapter on 'Moon Creatures') but also from the study of people, such as in 'Writing about People' and the comic sketches of relatives in his collection of poems in the chapter 'Meet My Folks'. He illustrates his points with examples from his own work as well as that of other writers, such as D. H. Lawrence, Emily Dickinson, Philip Larkin, Vasco Popa, and, of course, Sylvia Plath.

In the key chapter 'Learning to Think' Hughes provides the reader with examples to show the importance of close observation and thinking, as an integral part of education. More importantly though he reveals that the true significance of what lies inside inner life is the world 'of final reality, the world of memory, emotion, imagination, intelligence, and natural common sense, … like the heart beat' (*PM*: 57). This is also closely related to the sense of wholeness that lies in Hughes's healing quest in his children's writing; the idea that binary dualities such as nature/culture, human/non-human, child/adult must also be reconnected in order to be made whole and in this way become healed.

Amongst the other chapters, there are two on writing prose 'Writing Novels: Beginning' and 'Writing Novels: Carrying On'. The final chapter 'Words and Experience' was taken from a talk Hughes gave for an adult audience in the series 'Religion in Its Contemporary Context' broadcast on the BBC Home Service. The fact that Hughes had originally written the talk for an adult audience did not prove to be an impediment for the poet, and confirms again the idea of his writing 'within hearing' of both adults and children. The chapter talks about the power of words, experience and imagination, and to do so Hughes tells the story of a tramp who came to the door and asked for money, and then explains the difficulty of describing the experience of watching the tramp walk away.

Traditionally, in literature, the tramp is normally considered as a wanderer, and an outcast from society, so, in line with Hughes's shamanic journey, it would appear that the poet uses the tramp as a symbolic archetype. Hughes is, in fact, connecting the world of spirits and the world of the living. This blurred division between the world of the living and dead that often appear throughout Hughes's adult poetry seems to be a way for Hughes to connect both worlds: that of the spirits with that of the living. If we draw on Hughes's Jungian notion of the shaman as a healer who gained access to the underworld of the psyche in line with that of Eliade who considers the shaman as a 'medicine-man, priest and psychopomp; that is to say, he cures sickness, he directs the communal sacrifices and he escorts the dead to the other world' (1957: 87), then the tramp appears to take on the role of intermediating between those worlds. Moreover, since humans cannot reach that world, for Hughes shamanism is the only way of regaining that alienated consciousness and recuperating the lost animal/spiritual state, and in this sense the shaman, who takes on the form of the tramp here, is an intermediary between the worlds of spirits, gods and human beings. Hughes's interest in shamanism and the universal goddess of nature is also closely related to the healing process and the shamanic experience that the poet undergoes. As Hughes wrote:

> The essential experience of the Shaman is purely psychic, a magical flight to the goddess, and a return to worldliness with something divine, a cure, an inspired answer, some kind of blessing which everybody recognizes.… This fable is the poetic substance of which all poems draw – whether their authors are conscious of it or not.[14]

According to Eliade in *Shamanism* (1964), the undergoing of the psychological processes of death, dismemberment and resurrection are central to the shamanic experience. Given, however, that Hughes thought that most adults were already beyond saving, in most of his adult poetry the shaman undergoes death, and is dismembered, but appears to be unable to achieve the final stages of resurrection and recovery. This idea of dismemberment is present in most of Hughes's major adult works, and has been discussed at length by both Sagar (1983) and Skea (2001). Hughes believed that adults required a much more radical process to achieve redemption than children, who did not need to undergo the process of physical dismemberment or ego dissolution. Since Hughes genuinely believed that adults had a lot to learn from children, it is in his children's writing that Hughes best achieves the shaman's final stages of resurrection and recovery, which lead to the regeneration of the spirit. In his essay 'Concealed Energies' he refers to the 'visors' that adults wear, observing that

> Children's sensibility, and children's writing, have much to teach adults. Something in the way of a corrective, a reminder ... Preconceptions are already pressing, but they have not yet closed down, like a space helmet, over the entire head and face.
>
> (*WP*: 29)

Hughes held that these preconceptions, together with the 'space helmet' of science and technology, were, like a visor, responsible for making adults lose the ability to embrace new ideas and connect with their inner selves.

During the decade of the 1970s, as Diane Middlebrook has noted, the majority of his works for adults were 'tediously abstract and esoteric – even repellent' (2004: 242). Following the shaman's journey through the underworlds depicted in *Crow* (1970), *Prometheus on His Crag* (1973), *Cave Birds* (1975) and *Guadete* (1977), they all reflect the deep depression that Hughes had sunk into following the deaths first of Assia and Shura, 23 March 1969, and, in May, that of his mother Edith. Shortly after these tragic events, in a letter to Baskin, Hughes reflects the enormous pain he was going through at the time and the way it influenced his writing:

> Crow has dragged my life into its vortex and the quicker I get it finished the better ... but everything tries to frustrate it.... If I can get to the end I hope the end will redeem the rest somehow.[15]

It is interesting that Hughes should use the verb 'redeem' here, given that it endorses the fact that in his adult work redemption is never fully attainable or complete, unlike in many of his children's works. So whilst the poet's writing for adults in this period seems to articulate the mental cure that Hughes was struggling to achieve as an artist in his creative life, it would appear that in his children's work he best achieved the sense of redemption and catharsis, in accordance with his belief in the healing power of imagination and the fact that children were more open and receptive to the subconscious, as well as to nature.

Concerning the poet's attitude towards children, in a letter to Lissa Paul in April 1984, Hughes skilfully summarises the way he thinks grown-ups should talk with children about poetry, which he describes as 'simply the name we give to a certain kind of writing' (*LTH*: 483) and in which he confesses that his aim was 'to direct readers (listeners) towards certain faculties – inner concentration, inner listening and dependence on the spontaneous mind rather than on the calculating and remembering mind' (*LTH*: 483). Throughout *Poetry in the Making*, Hughes shows the importance he places on practice and experience, as fundamental tools in his technique to teach children how to write. As in 'Myth and Education' (1970), Hughes highlights the importance of education, and shows how, in his view, education should elevate 'experience' and 'feelings' over the abstract ways of learning.

In order to gain a better insight to Hughes's own ideas on the education of children, we might look at his first essay with the title 'Myth and Education' (1970),[16] which also develops the poet's true credo behind the working ideas of *Poetry in the Making*. In it Hughes also affirms his ideas and their centrality to his views on the education of children. Unlike the second version of 'Myth and Education', the first is written in a far more simple and critical tone. Hughes himself claims that it offers 'a few generalities behind my notion of writing stories for children with some sort of educational purpose' (1970: 55), whilst the second essay, collected in *Winter Pollen* (1994), offers a more distilled academic tone.

Hughes begins by making reference to Plato's notion that a good education for children ought to begin with traditional myths and tales, and that the importance of education is to build the child's imagination. Plato, he claims, 'wanted to create what we would also like to create, ideal citizens' (Hughes, 1970: 55). However, as the poet states, 'We can imagine what would happen if we proposed now that all education in England up to the school age of 11 be abolished and there be put in its place a huge system of storytelling' (Hughes, 1970: 55). According to Hughes the school curriculum is the 'outcome of three hundred years of rational enlightenment, which had begun by questioning superstitions and ended by prohibiting imagination itself as a reliable mental faculty' (1970: 55).

Hughes's attack on rational enlightenment serves as his argument that modern education, by inhibiting imagination, was destroying children's creative process. He claimed that a rationalistic, scientific mode of thinking dominated both our education system and our society and that the result was 'something resembling mental paralysis' (Hughes, 1970: 56) and argues explicitly here that this mode of thinking which has underlined the basis of modern education has threatened the creative process and the development of our imagination.

The importance of freeing imagination as a creative source of energy was essential to Hughes, since it played a vital role in his healing quest. As a result, in the second version of his essay 'Myth and Education', the poet gives undeniably more importance to imagination over education, whilst claiming that 'education neglects this faculty completely' (*WP*: 142). He discusses the role of imaginative literature and myth in restoring the harmony between our inner and outer worlds, two worlds 'which we have to live in simultaneously' (*WP*: 143). In both essays however Hughes expresses his concerns for an education system that neglects and inhibits imagination by blocking the development of the child's creative process and as Lissa Paul outlines 'he began to formulate the idea (learned from his knowledge of myth and folklore) that imaginative stories can act as antidotes, that they could potentially counteract the stultifying effects of modern life' (2005: 257).

Hughes's own ideas would appear to be very much in line with those resounding words of Charles Dickens in the opening chapter of *Hard Times* (1854), when the schoolmaster, Mr Gradgrind, drills into his audience the importance of facts: 'Now what I want is, Facts. Teach these boys and girls nothing but Facts. Facts alone are wanted in life. Plant nothing else, and root out everything else' (2006 [1854]: 3). Gradgrind makes it clear that his life is dictated by fact and that his 'model' school teaches nothing but facts, having eradicated all imaginative subjects from the curriculum. Dickens's criticism of dictatorial utilitarianism is not far off from Hughes's own criticism of the dangers of educating a society based on facts, at the expense of more imaginative pursuits. As Hughes warns us 'imagination isn't merely a surplus mental department meant for entertainment, but the most essential piece of machinery we have if we are going to live the lives of human beings' (1970: 57).

Plato is not only central to the poet's sense of imagination but also to his idea of education, not only because he believed, like Hughes, in the power of the word, but because Plato foresaw the despair of the human condition, and the passivity and detachment resulting from a society that failed to nurture the imagination as a mental faculty in young children. Plato, says Hughes, also put forth the idea that literature can make us emotionally and physically stronger and recognised the therapeutic power it had. In Hughes's words:

> He understood what the last three hundred years in Europe have forgotten, that without full operation of the various worlds and heavens and hells of imagination, men become sick, mechanical monsters.... Life becomes no longer worth living and the life instinct begins to fail.
>
> (1970: 60)

The sense of imagination related to that of the 'life instinct' that Hughes is referring to here underlines the connection between poetry and healing, and can be read in relation to our ecological crisis and the difficult task that lies ahead of the poet to recover what we have lost.

Hughes considered that children's literature was to take on new challenges and assume a new importance, without forgetting that storytelling and myths, should be the central point of education. He concludes his first essay by explaining the genesis of *The Iron Man* (which does not appear in his second and less personalised essay) and does so by showing us the way in which he uses certain symbols and the idea of myth to activate the child's inner world, that belong to our imagination.

The storyteller and the power of myth

In his second revised version of 'Myth and Education' Hughes explains how the myths and legends that Plato proposed as the ideal education for young children, can also be seen as 'negotiations between the powers of the inner world and the stubborn conditions of the outer world, under which ordinary men and women have to live' (*WP*: 151).

Children's fiction can be explored for archetypal plots, such as the recurring story of *The Hero with a Thousand Faces* (1949) that Joseph Campbell interprets according to the parameters of Jungian theory, and which he finds retold in myths from around the world.[17] In line with Jung, Hughes claims that myths contained cathartic healing properties that were able to restore the balance between our inner and outer world. In his second essay, he outlines how Plato's idea of education, in relation to storytelling and myth works on children: 'A child takes possession of a story as what might be called a unit of imagination' (*WP*: 139). And, as a result, 'The child can re-enter the story at will, look around him, find all those things and consider them at his leisure' (*WP*: 139). By defining a story as a 'unit of imagination', Hughes also echoes Claude Lévi-Strauss. Lévi-Strauss conceived myths as highly structured semiotic webs which provided the mind with patterns of psychological associations enabling one to think through abstract philosophical questions, on a universal level since 'a myth is still felt as a myth by any reader anywhere in the world. Its substance does not lie in its style, its original music, or its syntax, but in the *story* which it tells' (1963: 210).

Myths then, for Hughes, are stories that reconcile contradictions, and in line with the belief that Hughes has in the healing power of stories, great myths are those that reunite the inner and outer world and render them in balance. Child psychologist and critic Bruno Bettelheim has argued that myths and stories as metaphors are powerful vehicles for healing. His work has shown that fairy tales can help children work out their fears in symbolic forms, and claims that 'The fairy tale is therapeutic because the patient finds his own solutions, through contemplating what the story seems to imply about him and his inner conflicts at this moment in his life' (Bettelheim, 1977: 25). According to Bettelheim, a child's desire to listen to the same story over and over again can be explained by their necessity to believe the tale and integrate it into their own experience. Like Hughes, Bettelheim believes that the power of the story to heal is a basic tenet of Hindu medicine. Hughes, in line with Bettelheim, talks in his first article of 'Myth and Education' about the magical powers attributed to stories in primitive cultures. According to Hughes a person is sick when 'there is something in his mind which he refuses to face' (1970: 58), and the way to release that repression is 'within the framework of a fantasy. Once the fantasy has made connection with the demon and given him a role, the person feels cured' (1970: 58).

In his children's writing the poet attempts to heal the split between the rational and imaginative through his myths and storytelling. Nothing was more important to him, throughout his entire working life, than to connect with children. Hughes thought, like Rachel Carson, that 'if facts are the seeds that later produce knowledge and wisdom, then the emotions and the impressions of the senses are the fertile soil in which the seeds must grow' (1998: 56). Hughes, very much like Carson, questions where we are today in our understanding about the sense of wonder in young children, urging us to use our senses to explore nature whilst advocating a sense of respect and responsibility towards the natural world in his writing. As he declares in a brilliant passage from the second of his essays, echoing Carson's own words:

> Every new child is nature's chance to correct culture's error ... They want to escape the ugliness of the despiritualised world in which they see their parents imprisoned. And they are aware that this inner world we have rejected is not merely an inferno of depraved impulses and crazy explosions of embittered energy. Our real selves lie down there.
>
> (*WP*: 149)

Hughes's statement draws on one of the keystones of this book: the fact that he firmly believed children to be more open and receptive than adults. In this sense he wrote not as Cushman emphasises 'within hearing of children' but instead 'within hearing of adults'. In a letter to Lissa Paul he explains why: 'Now it seems to me, children don't demand anaesthetics. One can communicate with children in a simple and whole way – not because they're innocent, but because they're not yet defensive' (*LTH*: 482). This style of communication was what he called a lingua franca: 'a style of communication for which children are the specific audience, but which adults can overhear [if] they suspend their defences and listen – in a way secretly – as children' (*LTH*: 482). Hughes generally believed that adults had a lot to learn from children, and that by means of breaking down the barriers between the adult/child reader/listener they might overhear and share with the young child the ability to embrace new ideas and connect with their inner selves. He firmly believed in the need to awaken the imaginative and creative inner consciousness of the reader. In order for this to happen, the young child, through stimulus and by showing an intellectual and emotional openness, was allowing himself/herself to be touched by that powerful element that Hughes called 'grace'. One of the stimuli to awaken a child's consciousness is that of visual art. Given the importance that Hughes places on the role of art it can be considered as a key concept in his healing quest, especially in his children's work.

Hughes's appreciation of visual art[18]

Throughout his life Hughes collaborated with a number of artists, from Jim Downer in the 1950s, as can be seen in his book *Timmy the Tug* (2009), Fay Godwin whose beautiful black and white photographs capture the landscapes of the Calder Valley and accompany Hughes's poems in *Remains of Elmet* (1979), and Reg Lloyd who collaborated during the 1980s on three of his children's books: *What is the Truth?* (1984), *The Cat and the Cuckoo* (1987) and *The Mermaid's Purse* (1993). However, the artist with whom Hughes maintained the longest and indeed most fruitful collaboration was the American sculptor Leonard Baskin. Baskin taught art at Smith College, Northampton, Massachusetts where Sylvia Plath also taught English, and it was there that he met the British poet in 1958. Both shared not only an extensive working relationship together but they also developed an unlikely friendship. Son of a Lithuanian rabbi and a white Russian mother, Jewish mythology and Ancient Greek legend lay at the heart of Baskin's work. The artist later moved with his family to Lurley Manor, near Tiverton, Devon, where he lived from 1974 to 1983, in order to be closer in proximity to Hughes. The life-long friendship and creative artistic collaboration between the artist and the poet which encompassed some 30 books, would last until Hughes's death. Some of the most notable examples of their collaboration can be found in the limited editions of *Crow* (1973), *Selected Poems* (1973), the American edition of *Season Songs* (1975), *Cave Birds* (1975), *Moon-Whales and Other Moon Poems* (1976), *Under the North Star* (1981) and *Flowers and Insects* (1986).[19]

Given the importance of visual art to Hughes, what exactly did the poet consider as the essential role of art? Hughes's own appreciation on what an artist should be can be seen in a letter to his brother Gerald in 1952: 'Whenever an artist is transferring to paper he is wasting paint and fooling himself.... Any fool can become a mirror if he practises hard, but only half a dozen in an age can be artists' (*LTH*: 18). Both Plato in *The Republic* and Aristotle in *The Poetics* agree that art is mimetic, that is, it is an imitation of the world. However, both Hughes and Baskin go beyond the idea of mimetics and myth, since for them art was not supposed to be beautiful, but rather stark. It was not about imitating reality but rather commenting on it. In this sense both the poet and the artist share a similar quality: they both disfigure and chip away at the outer layers of our material world in order to reveal the raw core and essence that lies within. Indeed, Baskin's work challenges the idea that art must be 'beautiful' to have value. According to Hughes, in his essay on Baskin 'The Hanged Man and the Dragonfly',[20] 'Artistic form is, by definition, a lens' (*WP*: 84). Baskin's 'Hanged Man' depicts an anonymous figure who could be a victim or a criminal. The figure became a leitmotif for the artist, reappearing throughout his career in different forms: prints, drawings and different sized bronze sculptures. Baskin's work encapsulates the artist's mind: his commitment to social justice and his ancillary faith in the redemptive power of art. Baskin's preoccupation with death can also be seen in his prints and drawings which reflect the image of man as a distorted figure representing the spiritual corruption of Western society. For Hughes the 'Hanged Man' is also the image of what he calls *mana*, that is, in Hughes's words, 'the body's natural response to grave hurt – a healing' (*WP*: 93). Whilst Baskin's hanged man symbolises pain and agony, the dragonfly depicts the transformation that it has undergone, and with it redemption.

Hughes's essay on Baskin reflects much of the poet's own views of art and aesthetics. In his words, 'Baskin writes, somewhere, that his subject is the wound. One could as truly say that his subject is *mana*. His real subject is the healing of the wound' (*WP*: 95). Both the poet and the artist project a similar sense of healing: art was as therapeutic and necessary for Baskin, as writing was cathartic for Hughes. In the same way, and despite its somewhat depressing subject matter, Baskin's art was essentially optimistic, given that he felt that the only way to triumph over the horrors of modern life was to address them directly. As Baskin professed:

> Have we, as twentieth-century humans, become so dehumanized that we can only participate in the art of the past as appreciators and not as creators? … he [Man] has charred the earth and befouled the heavens more wantonly than ever before. He has made of Arden a landscape of Death. In this landscape we dwell, and with these images we must live.
>
> (1961: 76)

In reality Baskin's words echo Hughes's own ongoing preoccupation with the dehumanisation of human society and the consequent alienation from nature. For both the artist and the poet, art and writing were, more than a creative process, in essence an attempt to reconnect binary dualities and heal the wounds, creating through their collaboration, and the close friendship that often resulted in a perfect symbiosis between the visual and the written.

Integrating theory with dialogue: bringing the stories back to earth

Lawrence Buell first coined the term *environmental imagination* (1995) with Henry David Thoreau in mind when he brought it to the forefront of ecocritical discussion. In his celebrated book Buell describes the book's three aims: '[a] broad study of environmental perception, the place of nature in the history of Western thought and the consequences for literary scholarship and indeed for humanistic thought in general of attempting to imagine a more "ecocentric" way of being' (1995: 1) and claims that he 'could not discuss green writing without relating it to green thinking and green reading' (1995: 1). The term *environmental imagination* however remains somewhat problematic to define, given that it does not lend itself to a specific definition, but can be easily recognised, in words of Joseph Meeker, in writers such as John Muir or Henry David Thoreau. According to Meeker 'What such writers share is a profound love of the natural world and an active curiosity about its complex processes' (2003: 77).

In relation with Hughes's children's writing, the term *environmental imagination* is particularly useful, drawing on Buell's sense of imaginative engagement with environmental themes, and the fact that what we write arises from the blueprints of our imagination, which, in turn, is formed by who we are. In

Hughes's children's writing, the creative process is strongly imprinted and influenced by aspects from both the artist's memory and experience, such as place, landscape, nature, ecology and history amongst others. Therefore, his early sense of place, his studies of anthropology, his affinity for animals and his preoccupation for environmental issues can all be considered as important elements of his own *environmental imagination*. Moreover, Hughes, anticipating Buell, firmly believed that the environmental crisis 'involves a crisis of the imagination the amelioration of which depends on finding better ways of imagining nature and humanity's relation to it' (Buell, 1995: 2). In his attempt to engage children's literature with our collective and individual imagination the poet was also attempting to influence their sense of environmental ethics.

The term *environmental imagination* can also be read in relation to the critic Northrop Frye's use of the term 'educated imagination' (1964), which he draws on to enquire about the social value of the study of literature, and the place imagination has in the learning process. According to the Canadian critic, literature *is* the world of our imagination, and belongs to the world we construct, and not the world we see. Frye's sense of literature, very much in line with that of Hughes, is closely associated with Jung's idea of archetypes and is related with primitive human experiences and myths. Frye sees literature as a whole, having built up its own internal archetypal images and patterns of organisation over the centuries. According to Frye, a literary archetype is 'a symbol, that is, a unit of a work of literary art, which connects one poem with another and thereby helps to unify and integrate our literary experience' (1957: 99). In the case of Hughes's children's writing, Frye's interpretation of the term 'educated imagination' together with Jung and Lévi-Strauss's interpretation of myths can be helpful for reading the writings of Hughes in order to illuminate Hughes's own sense of *environmental imagination* and acknowledge its roots.

According to the *Encyclopaedia of Environmental Ethics and Philosophy* (Callicott and Frodeman, 2009), *environmental education* emerged as a subfield in American curriculum studies in the wake of David Orr's *Ecological Literacy* (1992). As a discipline of the curriculum, *environmental education* has moved closer to mainstream education in other parts of Europe, thanks to The Foundation for Environmental Education, based in Denmark, a 'non-government, non-profit organisation promoting sustainable development through environmental education'[21] which started in 1981 and which now has over 60 members. Simply put, *environmental education* is therefore a learning process that increases people's knowledge and awareness about the environment. However, as Joy Palmer has suggested, 'The evolution of environmental education has incorporated the significant influence of some of the "great" eighteenth – and nineteenth – century thinkers, writers and educators, notably Goethe, Rousseau, Humboldt, Haeckel, Froebel, Dewey and Montessori' (1998: 4). The fact that Palmer considers Rousseau together with other educators as pioneers in a pre-environmental movement thinking suggests a link between the values of education with the importance of environment as in Hughes's thinking.

It is precisely this link with education and, drawing on Rachel Carson's sense of 'environmental awareness' related to instilling in the young child an appreciation of the natural world, that comes to mind when using the term *environmental education* in relation to Hughes. Furthermore, given that ecocriticism and children's literature often share a similar concern related to a particular consciousness of the importance of *environmental education* which begins during the early years of childhood, *environmental education* should be a central issue to both children's literature and ecocriticism.

Following both Carson and Hughes, it is clear that the stories that we read as children influence the way we respect and value nature later in our lives. Children's writers, therefore, are in a position of singular responsibility, for it is largely up to them to transmit not only cultural values, but to educate and teach respect for our natural world and to invite young readers into the negotiation of responsibility with the other. Hughes, like Carson, believed that children needed direct contact with nature, and understood the importance of bonding with nature in those early years, for it is, as Carson states, in those years of wonder when the child's innate curiosity for the environment is shaped:

> The years of early childhood are the time to prepare the soil. Once the emotions have been aroused – a sense of the beautiful, the excitement of the new and the unknown, a feeling of sympathy, pity, admiration or love – then we wish for knowledge about the object of our emotional response.
> (1998: 56)

Carson's sense of 'wonder' then appears to be very much like Hughes's own sense of 'grace' and the importance he gave to engaging with children throughout his working life in order to develop their creativity and imagination. Hughes developed his sense of environmental awareness through practical activity and aspired to teach children to respect the natural world. He did this through his own call to action and through the importance he placed on what could be called his sense of *environmental education*.[22] Hughes's *environmental education* is therefore closely related to Carson's sense of 'environmental awareness' in his attitude towards developing awareness in children through practical activities and conservation issues, together with the importance he places on the values of education in young children in creating responsible attitudes and promoting the development of an environmental ethic. For Hughes the key to successfully achieving his own shamanic healing quest lies in part of an ongoing venture which he carries out to reconnect dualities. These dualities are not just hierarchical dualisms of binary pairs (male/female, culture/nature, self/other, human/non-human) but take account of other dualities, such as that of the child/adult reader.

Hughes's writing for children draws on a Wordsworthian view of nature, the idea that as children we are in perfect harmony with nature, but as we grow older we lose that harmony and develop different barriers to protect the self, so that as adults we have lost that sense of unity. The poet, by reintroducing and connecting dualities, is attempting not only to speak to children, but also to those that are 'within hearing' (Cushman, 1983: 255), that is to the gatekeepers of children: to both the parents and their teachers. As a writer he was not inclined to talk down to children or deliver them 'anaesthetics' to cope with the matters of life and death. Moreover, by looking at the interconnections that lie between nature and culture Hughes holds in question not only our concept of 'nature' but also our 'sense of place' and the importance of our surroundings. Like Carson, and anticipating Buell, he focuses on 'the more delicate issue of how the sense of place can be kept alert and sensitive' (Buell, 1995: 261), and subscribes to the belief that the interconnectivity between people and their own particular local bioregion holds a certain meaningfulness and value which can contribute towards environmental ethics on a wider and more global scale.

For Hughes literature acts as a process of healing the 'disenchantment' of human beings and in doing so he follows Carson by recreating that sense of wonder that humans have lost through alienation from nature. Hughes's adult poetry looks at what human nature is made of, dismembering the parts, whilst his children's poetry recovers and *reconnects us* (both the adult and the child) to Carson's sense of wonder. In *Reconnecting with John Muir* (2006), Gifford states that, 'the teacher is the writer, is the critic, is the scholar, is the human being, is the neighbour of other organisms, is a small part of our biosphere home. Only reconnect' (16). This idea of reconnecting seems particularly appropriate to use in relation to Hughes, where his personal experience and his writing is fired by a love of literature and the natural world. However, Hughes is not just an observer of nature, but a participant, and moreover, his writing requires the reader to become 'active'. Hughes speaks to our imagination. He writes about education, culture, imagination and nature as things he cares for, and by being part of that project he is *reconnecting* and bringing his stories back to earth. Given that Hughes's writing looks at these interconnections, terms such as *environmental imagination* and *environmental education* can prove helpful to understand how the poet describes the therapeutic value of human interaction with nature. These concepts are particularly useful to show how Hughes's writing for children puts into practice the ecological principle of interconnectedness and at the same time is part of a project, as one might say, to bring the stories he tells 'back to earth'.

Notes

1 Sagar's pamphlet *Ted Hughes*, originally published by Longman, 1972, was the first short monograph on Hughes, and was later enlarged as *The Art of Ted Hughes*, Cambridge University Press, 1978.

2 Paul's essays include 'The Return of the Iron Man', *The Horn Book Magazine*, 76:2 (March/April 2000), pp. 218–225; '"Writing Poetry for Children is a Curious Occupation": Ted Hughes and Sylvia Plath', *The Horn Book Magazine*, 81:3 (May/June 2005), pp. 257–267; 'Ted Hughes and the Old Age of Childhood', in *Poetry and Childhood* (2010: 33–44).

3 See *LTH*: 481–484.

4 Other essays by Kazzer include 'Difficulties of a Bridegroom', in Bertrand Rougé (ed.), *Q/W/E/R/T/Y: Littératures et Civilisations des Pays Anglophones* (Pau: Publications de l'Université de Pau, 9 October 1999), pp. 187–201 and 'Family Relations – Traces of a Cosmology in Ted Hughes's Creation Tales', published online at www.earth-moon.org, July 2010.
5 My thanks to Claas Kazzer for allowing me to read his insightful article '"Earth-Moon": Ted Hughes' Books for Children (and Adults)', written in the form of a paper given for a Ted Hughes conference at Lyon, 2000, which unfortunately remains unpublished.
6 See Cook and Gowar (2011) and Gowar (2011).
7 *The Ted Hughes Society Journal* is an online, peer-reviewed scholarly publication that was created following the 2010 Ted Hughes International Conference 'From Cambridge to *Collected*' held at Pembroke College, Cambridge. See: www.thetedhughessociety.org/.
8 Published as *The Iron Man* (1968) in the UK. In America the title was changed to *The Iron Giant* so as to avoid confusion with Marvel Comics' character Iron Man.
9 Having recently discussed this with Lissa Paul she has assured me that credit for inclusion of *The Iron Man* goes to Zipes and that she actually wanted all of *What is the Truth?* However, due to editorial decisions to include no more than one work by any given author it made more sense contextually to include *The Iron Man.*
10 The popularity of Plath's work in America, especially after her death, resulted in a backlash against Hughes. Emerging feminists saw Hughes as 'the murderer of Plath' to the point that his name was chiselled off Plath's gravestone. The hecklings that he suffered at public events became so intense that it caused Hughes to put an end to his public readings in America and Australia.
11 See Gadd *et al.* (2014).
12 Quoted here in Sagar (2006). Original quote comes from Plath's *Journals*. See Hughes and McCullough (1982).
13 The background behind the poem is discussed in detail in Chapter 1 (this volume).
14 MSS 644, Box 115, FF20, Robert W. Woodruff Library, Emory University, Atlanta, Georgia, USA.
15 Letter dated 16 July 1969, Add MS 83684/40, The British Library, London.
16 Hughes first essay entitled 'Myth and Education' was published in *Children's Literature in Education*, I (March, 1970), pp. 55–70. A different revised version appeared in *Writers Critics and Children* (Fox *et al.*, 1976), which is the same as that collected in *Winter Pollen* (1994).
17 Campbell applies Jung's concepts of the collective unconscious and archetypes to his study of mythology, recognising the similar underlying themes that appeared in all stories and myths, regardless of the storyteller's location, religious beliefs or socio-economic factors in what he termed the 'Monomyth', which shows the structure of many heroic myths and can be traced in Hughes's own storytelling as part of his shamanic healing quest. Campbell distinguishes tree sections: that of the Departure, the Initiation and the Return, coinciding with the same pattern that many traditional fairy tales and folk stories follow.
18 For a more detailed study of Hughes's collaborations with artists see Kerslake (forthcoming 2018).
19 Hughes's collaborative work with Baskin, as well as that with Reg Lloyd, is discussed in Chapter 5 to assess whether the drawings and illustrations that accompany the poet's work successfully achieve the creative synthesis of the written word together with the visual image.
20 Hughes's essay 'The Hanged Man and the Dragonfly' which he wrote as an introduction to *The Complete Prints of Leonard Baskin*, 1984, was collected in *Winter Pollen* (1994).

21 For more information see their web page: www.fee-international.org/en.
22 Clear examples of this can be seen in Hughes's involvement in numerous educational and environmental projects such as the tree planting scheme he set up with local schools in Devon in 1972, his involvement with Michael Morpurgo in setting up Farms for City Children, in 1976, and the Sacred Earth Drama Trust, founded by Hughes in the form of a playwriting competition for children in the 1990s, which aimed to make children spokespeople for the environment.

Archive material

Add MS 83684/40. The British Library, London.
MSS 644, Box 115, FF20. Unidentified, Series 2: Writings by Ted Hughes, circa 1949–1990, Subseries 2.3d: Essays, Reviews, Talks from Ted Hughes Papers, 1940–1999 (Robert W. Woodruff Library, Emory University, Atlanta, Georgia, USA).

References

Baskin, Leonard, 'The Necessity for the Image', *Atlantic Monthly* (April 1961), pp. 73–76.

Bate, Jonathan, *Ted Hughes: The Unauthorised Life* (London: William Collins, 2015).

Bettelheim, Bruno, *The Uses of Enchantment* (New York: Vintage Books, 1977).

Bradshaw, Graham, 'Ted Hughes' *Crow* as Trickster Hero', in P. V. A. Williams (ed.), *The Fool and the Trickster* (Cambridge: Brewer/Rowman & Littlefield, 1979), pp. 83–108.

Buell, Lawrence, *The Environmental Imagination: Thoreau, Nature Writing and the Formation of American Culture* (Cambridge: Harvard University Press, 1995).

Callicott, J. Baird and Robert Frodeman (eds), *Encyclopedia of Environmental Ethics and Philosophy* (New York: Macmillan, 2009).

Campbell, Joseph, *The Hero with a Thousand Faces* (Princeton NJ: Bollingen Foundation, 1949).

Carson, Rachel, *The Sense of Wonder* (New York: Harper & Row, 1965; New York: HarperCollins, 1998).

Cook, Peter, 'Poets in the Making: Ted Hughes, Poetry and Children', in Morag Styles, Louise Joy and David Whitley (eds), *Poetry and Childhood* (Stoke-on-Trent: Trentham Books, 2010), pp. 195–201.

Cook, Peter and Mick Gowar, 'Within the Hearing, and Seeing, of Children', *The Ted Hughes Society Journal*, 1 (Summer 2011), pp. 32–38.

Cushman, Keith, 'Hughes' Poetry for Children', in Keith Sagar (ed.), *The Achievement of Ted Hughes* (Manchester: Manchester University Press, 1983), pp. 239–256.

Dickens, Charles, *Hard Times* (London: Elibron Classics Series, 2006 [1854]).

Eliade, Mircea, *Myths, Dreams and Mysteries* (London: Harvill Press, 1957).

Eliade, Mircea, *Shamanism* (London: Routledge, 1964).

Faas, Ekbert, *Ted Hughes: The Unaccommodated Universe* (Santa Barbara, CA: Black Sparrow Press, 1980).

Fox, Geoff, Graham Hammond, Terry Jones, Frederic Smith ad Kenneth Sterck (eds), *Writers, Critics, and Children: Articles from Children's Literature in Education* (New York and London: Heineman Educational Books, 1976).

Frye, Northrop, *Anatomy of Criticism: Four Essays* (Princeton: Princeton University Press, 1957).

Gadd, Ian, Terry Gifford and Lorraine Kerslake, 'A Note on Ted Hughes and Jonathan Swift', *Ted Hughes Society Journal*, 4:1 (2014), pp. 85–88.

Gifford, Terry, *Reconnecting with John Muir: Essays in Post-Pastoral Practice* (Athens: University of Georgia Press, 2006).

Gifford, Terry and Neil Roberts, *Ted Hughes: A Critical Study* (London: Faber & Faber, 1981).

Gowar, Mick, 'Poetry in the Making: Fifty Years Old', *Ted Hughes Society Journal*, 2 (Winter 2011), pp. 1–13.

Hirschberg, Stuart, *Myth in the Poetry of Ted Hughes* (Dublin: Wolfhound Press, 1981).

Hughes, Ted, *Meet My Folks* (London: Faber & Faber, 1961).

Hughes, Ted, *The Earth-Owl and Other Moon-People* (London: Faber & Faber, 1963).

Hughes, Ted, *Poetry in the Making* (London: Faber & Faber, 1967).

Hughes, Ted, *The Iron Man* (London: Faber & Faber, 1968).

Hughes, Ted, 'Myth and Education', *Children's Literature in Education*, I (March 1970), pp. 55–70.

Hughes, Ted, *Crow: From the Life and Songs of the Crow* (London: Faber & Faber, 1970).

Hughes, Ted, *Prometheus on His Crag* (London: The Rainbow Press, 1973).

Hughes, Ted, *Selected Poems 1957–1967* (New York: Harper & Row, first American edition, 1973).

Hughes, Ted, *Cave Birds* (Ilkley: Scolar Press, 1975).

Hughes, Ted, *Season Songs* (New York: Viking, 1975).

Hughes, Ted, *Season Songs* (London: Faber & Faber, 1976).

Hughes, Ted, *Moon-Whales and Other Moon Poems* (New York: Viking, 1976).

Hughes, Ted, *Gaudete* (London: Faber & Faber, 1977).

Hughes, Ted, *Cave Birds* (London: Faber & Faber, 1978).

Hughes, Ted, *Remains of Elmet* (London: Faber & Faber, 1979).

Hughes, Ted, *Under the North Star* (London: Faber & Faber, 1981).

Hughes, Ted, *What is the Truth?* (London: Faber & Faber, 1984).

Hughes, Ted, 'The Hanged Man and the Dragonfly', Introduction to *The Complete Prints of Leonard Baskin: A Catalogue Raisonné, 1948-1983*, dirs. Alan Fern and Judith O'Sullivan (Boston: Little, Brown and Company, 1984), p. 19.

Hughes, Ted, *Flowers and Insects* (London: Faber & Faber, 1986).

Hughes, Ted, *The Cat and The Cuckoo* (London: Sunstone Press, 1987).

Hughes, Ted, *The Iron Woman* (London: Faber & Faber, 1993).

Hughes, Ted, *The Mermaid's Purse*, Illus. R. J. Lloyd (Bideford: Sunstroke Press, 1993).

Hughes, Ted, *Winter Pollen* (London: Faber & Faber, 1994).

Hughes, Ted, *Difficulties of a Bridegroom* (London: Faber & Faber, 1995).

Hughes, Ted and Frances McCullough (eds), *The Journals of Sylvia Plath* (New York: Dial Press, 1982).

Kazzer, Claas, 'Difficulties of a Bridegroom', in Bertrand Rougé (ed.), *Q/W/E/R/T/Y: Littératures et Civilisations des Pays Anglophones* (Pau: Publications de l'Université de Pau, 9 October 1999), pp. 187–201.

Kazzer, Claas, 'What is the Truth? Ted Hughes, Childhood, Memories and Stories', 2010. Web. 24 March 2012. www.ted-hughes.info/criticism/online-articles/kazzer-claas/what-is-the-truth.html.

Kerslake, Lorraine, 'Hughes and His Collaborations with Artists', in Terry Gifford (ed.), *Hughes in Context* (Cambridge: Cambridge University Press, forthcoming 2018).

Lévi-Strauss, Claude, 'The Structural Study of Myth', in *Structural Anthropology* (London: Allen Lane, 1963), pp. 202–212.

Lockwood, Michael, 'Ted Hughes: The Development of a Children's Poet', *Children's Literature in Education*, 40:4 (December 2009), pp. 296–305.

Meeker, Joseph, 'Willa Cather: The Plow and the Pen', in Susan J. Rosowski (ed.), *Willa Cather's Ecological Imagination* (Cather Studies Volume 5. Lincoln, NE: University of Nebraska Press, 2003), pp. 77–88.

Middlebrook, Diane, *Her Husband* (London: Little, Brown, 2004).

Morrison, Blake, 'Man of Mettle', *The Independent on Sunday, Review* (5 September 1993), pp. 32–34.

Orr, David, Ecological Literacy. Education and the Transition to a Postmodern World (Albany: State University of New York Press, 1992).

Palmer, Joy A., *Environmental Education in the 21st Century: Theory, Practice, Progress and Promise* (London and New York: Routledge, 1998).

Paul, Lissa, 'The Return of the Iron Man', *The Horn Book Magazine*, 76:2 (March/April 2000), pp. 218–225.

Paul, Lissa, '"Writing Poetry for Children is a Curious Occupation": Ted Hughes and Sylvia Plath', *The Horn Book Magazine*, 81:3 (May/June 2005), pp. 257–267.

Paul, Lissa, 'Ted Hughes and the Old Age of Childhood', in Morag Styles, Louise Joy and David Whitley (eds), *Poetry and Childhood* (Stoke-on-Trent: Trentham Books, 2010), pp. 33–44.

Philip, Neil, 'The Shawl of the Beauty of the World: The Children's Books of Ted Hughes', *Signal: Approaches to Children's Books*, 100 (2003), pp. 191–202.

Reid, Christopher (ed.), *Letters of Ted Hughes* (London: Faber & Faber, 2007).

Roberts, Neil, *Ted Hughes: A Literary Life* (Basingstoke: Palgrave Macmillan, 2006).

Sagar, Keith (ed.), *Ted Hughes* (Harlow: Longman for the British Council, 1972).

Sagar, Keith (ed.), *The Art of Ted Hughes* (Cambridge: Cambridge University Press, 1978).

Sagar, Keith (ed.), *The Achievement of Ted Hughes* (Manchester: Manchester University Press, 1983).

Sagar, Keith (ed.), *The Laughter of Foxes* (Liverpool: Liverpool University Press, 2006).

Skea, Ann, 'Poetry and Magic – The Hierophant', 2001. Web. 15 May 2014. http://ann.skea.com/Hieroph.htm.

Styles, Morag, *From the Garden to the Street: Three Hundred Years of Poetry for Children* (London: Cassell, 1998).

Styles, Morag, Louise Joy and David Whitley (eds), *Poetry and Childhood* (Stoke-on-Trent: Trentham Books, 2010).

Whitley, David, '"Imaginary Gardens with Real Toads in Them": Animals in Children's Poetry', in Morag Styles, Louise Joy and David Whitley (eds), *Poetry and Childhood* (Stoke-on-Trent and Sterling, USA: Trentham, 2010), pp. 187–194.

Whitley, David, '"The Fox is a Jolly Farmer and We Farm the Same Land": Ted Hughes and Farming', in Mark Wormald, Neil Roberts and Terry Gifford (eds), *Ted Hughes: From Cambridge to Collected* (Basingstoke: Palgrave Macmillan, 2013), pp. 96–111.

Wilson, Anthony, 'Ted Hughes's Poetry for Children', *Children's Literature in Education*, 32:2 (2001), pp. 77–90.

Wormald, Mark, Neil Roberts and Terry Gifford (eds), *Ted Hughes: From Cambridge to Collected* (Basingstoke: Palgrave Macmillan, 2013).

4 Hughes's children's plays

Ever since ancient times, drama has played a crucial role within society. The theatre was both a social and 'spiritual' gathering; one in which the members of a community got together to listen to and participate in the stories that gave meaning to their existence. The stories told through the voice of the storyteller/shaman provided in this way, a vehicle for the audience to enter into the world of myths. Hughes's children's plays draw on this traditional sense of drama, as a ritual and symbolic narrative, and one which gives importance to the role of imagination, so that drama becomes not only a way to enact myths, but also, and especially in the case of the poet, one in which to discover the world of truths. Hughes's own healing quest is intrinsically related to his quest for truth, especially in the realm of his writing for children, which reveals itself as a truth-revealing process to enable the young reader to establish contact with the world-as-it-really-is. He problematises his version of the 'truth', a key element in his children's writing, understood as something relative and not absolute, and uses it in a playful sense for his child reader/listener.

Hughes's children's plays reveal how the role of theatre is linked to that of imagination as an ecological force which can nurture our sense of wonder and enact the poet's own belief in the power of poetic imagination as an instrument for reaching the truth and as a way to reintegrate our inner and outer worlds. In words of Theresa May: 'Theatre functions as a field of exchange where stories take flight, moving between the permeable spheres of self and community, then out into the terrain of lived experience' (2005: 86). Drawing on May's sense of 'ecodramaturgy',[1] the term 'ecodrama' can be used to look at the ecological implications – including the human/non-human relationships and conflicts underlying nature and culture – that are revealed when Hughes's children's plays are read through the lens of ecocriticism.

In line with the poet's shamanic quest, and given the importance of his representation of the goddess as nature, it is in his children's work, and more precisely in his early playwriting, where Hughes's sense of ecocritical awareness can be seen as first developing. Certainly it is in his initial radio plays where he first enquires what happens if the balance between the energies of the universe is not fulfilled and exposes the dualities that have disconnected human beings from nature throughout Western Christian culture.

Hughes's distrust of the rational is strongly linked to his belief in imagination as a tool of psychic regeneration and his own shamanic healing quest. Given the convictions that underlie his poetic voice and the role he took on to communicate and heal human alienation from nature, his children's plays often represent conflicts underlying different dualisms such as that of nature and culture, whilst highlighting the importance of being attuned to the natural world. Many of his collected plays reflect ecological debates, and can offer suggestive ways to explore ecocritical questions within the field of 'ecodrama', given the way in which Hughes as a storyteller and playwright not only *writes* his plays, but engages or reflects environmental issues through the reading of the performance itself. These are, of course, the same dualities that will be reconnected more successfully in *Orpheus* and in his later poetic writing in order to heal the rift that has alienated human beings from nature.

(Reading) Shakespeare and the goddess: Hughes's initiation in drama

Hidden between the pages of a 1947 edition of the Mexborough local newspapers lies a curious black and white photo of a young Ted Hughes as a sixth-former. In it he plays the lead role of David Charleston, the lighthouse keeper, in a stage performance of Robert Ardrey's *Thunder Rock* produced by Pauline Mayne, Hughes's English teacher at Mexborough Grammar School.[2] According to the review published in the school magazine *The Don and Dearne* (*Being the Magazine of Mexborough Grammar School* as it boldly claims in elaborate font) in July 1947, he was 'if anything, too restrained, so that at times his words were difficult to hear. Nevertheless, his stage presence and sense of movement and gesture compensated for this flaw' (1947: 13–14).

For the following two years Hughes co-wrote, cast and directed the sixth form dramatic revues.[3] Hughes was in fact joint sub-editor of these revues with his sister Olwyn, supervised by his mentor John Fisher, who was, at the time, editor. Although the poet's own interest in performing did not go beyond his school days, it was at Mexborough Grammar School, and largely motivated by his English teachers Pauline Mayne and John Fisher, that he took his first steps on stage and developed his ongoing interest in drama.[4] Hughes's first real connection to drama however, came through encouragement from Olwyn, who was at the time his literary mentor, to read Shakespeare. From then on he read and reread Shakespeare throughout his life. Between 1949–1951, whilst doing his National Service, stationed at an isolated radar station in the Yorkshire moors, Hughes declared that he had 'nothing to do but read and re-read Shakespeare and watch the grass grow' (Sagar, 1975: 8). Later, in 1957, in a letter to Lucas Myers, written whilst in Massachusetts, he claims, in what he calls 'a deliberate project' (*LTH*: 105), to have reread all of Shakespeare following what he considered their original order of composition. Hughes's ongoing obsessive fascination with Shakespeare climaxed in his monumental work *Shakespeare and the Goddess of Complete Being* (1992), a mission that consumed him for over 20 years, to the point that, as Hughes himself put it, it was 'destroying his immune system' (*LTH*: 719) almost until his dying day. In his project the poet's basic underlying thesis is that Shakespeare had a myth of his own. His epic piece of literary criticism, of intrinsic complexity and length, attempts to unravel the truth behind the underlying connections between the myths, plots and imagery of Shakespeare's playwriting.

According to Hughes, all of Shakespeare's works clearly manifest the underlying divine or demoniac pattern found in what he calls Shakespeare's 'Tragic Equation' (*SGCB*: I), derived from the poems 'Venus and Adonis' and 'The Rape of Lucrece'. It is here that Sagar claims that Hughes's fullest description of the myth of the goddess, which accounts for his 'tragic equation', can be found. In it Hughes states that we cannot escape the powerful forces of the goddess, since we are both a part of nature despite remaining apart from it. The goddess cannot be divided. Therefore, when confronting the Goddess

> the ego's extreme alternatives are either to reject her and attempt to live an independent, rational, secular life or to abnegate the ego and embrace her love with 'total, unconditional love', which means to become a saint, a holy idiot, possessed by the Divine Love.
>
> (*SGCB*: 392)

According to Hughes, man will always choose the former and, by doing so, he rejects the Goddess, leading in the first phase of the tragedy, to his own self-destruction.

For Hughes, like Shakespeare, the divine mother, the sacred bride and the queen of hell are all the same woman. In his tragic equation Hughes claims that Shakespeare's male characters simply cannot engage successfully and rise above the threefold muse. The tragedy comes when the characters struggle, always unsuccessfully and often accompanied by violence, to accept the very same women who inspired their adoration. This is, of course, the same goddess that the poet took from Graves's *The White Goddess* (1948), which he had received in his teenage years as a present from his English teacher, John Fisher, and which he later brought to his own reading of Shakespeare. More importantly however, it is also the same multi-faceted goddess who appears in different forms as mother, bride and hag in Hughes's poetry.

Hughes's own underlying interest in both shamanism and classical mythology, that can be traced throughout the entire body of his work, can also be linked to his reading and writing on Shakespeare. For Hughes, Shakespeare is something of a totem poet. Given that Hughes regarded the poet as both shaman and prophet, he describes Shakespeare's shamanic task to communicate through the act of transforming experience into art. In words of Hughes: 'Art is a psychological component of the auto-immune system that gives expression to the healing process. That is why great works of art make us feel good' (Sagar, 2006: xi). The poet's generic use of the word 'artist' here can be read in close association to his own healing quest, referring to his assumption that the poet/artist 'works' for the rest of society as a kind of 'psychic healer' whose role coincides with that of the shaman. Despite the fact that shamanism is the oldest form of mind/body healing known to human beings, in the framework of ecopsychology, it can be considered as an effective context for restoring the balance and reconnecting our severed selves to nature. In the case of Hughes, his own shamanic healing quest takes place throughout his creative work, and can be seen here in the progression of his playwriting techniques, and his attempt to reconnect the balance between culture – in the form of human consciousness – and nature.

Working for the British Broadcasting Corporation (BBC): a career in playwriting

During his career, together with his poetry and prose, Hughes produced many plays both for the radio and for the stage. Despite the fact that both Hughes and Plath were keen theatre goers, his dramatic work, both for children and for adults, shows little resemblance to the gritty social commentary of post-war English society that dominated the London stage of his own time.[5] Hughes wrote over 18 plays for both children and adults, and worked on a variety of other drama projects throughout his life, from his collaboration with Peter Brook on a stage adaptation of Seneca's *Oedipus* for the National Theatre Company in 1969 up to 'The Sacred Earth Drama Trust' founded by Hughes in the 1990s to foster environmental awareness through creativity in the form of a playwriting competition for young children which was to be published in a collection and used in schools. In the foreword to the collection Hughes draws on a crucial point related to the effect of theatre on the intended audience: the fact that 'adults, parents and teachers watch children's drama through the eyes of the child within themselves. It reaches them in ways they never can be reached if addressed simply as adults' (1993: vii–viii).

In the same way, Hughes also remarked on the importance of letting down the barriers between the adult/child reader given the fact that adults 'are more affected and moved by children's drama, which they watch through the eyes of their children, than they are by adult drama, which they watch through their visors' (*LTH*: 577). These would appear to be the same visors that have repressed human emotion, distorting the reality and alienating us from nature. What also emerges here is the idea of writing in a 'lingua franca', or as Hughes puts it 'a style of communication for which children are the specific audience, but which adults can overhear' (*LTH*: 482).[6]

Despite his eclectic range of drama, critics have often overlooked the importance of Hughes's early children's plays, most of which had been commissioned as radio plays for the BBC Schools English radio series 'Listening and Writing' and 'Living Language' in the 1960s. As has been previously mentioned, it was between 1960 and 1970 that Hughes's main works in the field of children's literature were conceived. Furthermore, for an important period of time, whilst Plath and Hughes were struggling to make ends meet, Hughes found in playwriting an attractive financial opportunity which was soon to become both his and Plath's main source of income. As Plath claims in a letter to her mother, dated 26 October, 1960, 'Ted's income from the BBC this year has been as good as a salary – we've about $1,600 in the bank here from our English writing' (1976: 398), and a year later writes that 'Ted's children's broadcasts on the BBC have been very enthusiastically received and he has an open ticket there to do as much as he wants' (1976: 423).

The first of Hughes's radio plays was 'The House of Aries', broadcast in November 1961. Hughes considered, however, that the play should never have been acted, due to the lack of action, character roles and argument. As he recognised it simply didn't work: 'I wrote it hastily, even casually, without thinking out its implications.... Altogether, I was a bit horrified when I saw what I had let out in the public' (*LTH*: 174). This comment highlights an important feature common to all of Hughes's children's plays: the fact that nearly all the plays were made as radio plays, and were not intended to be put on stage. Hughes's plays are created through the idea of performance not in the traditional sense but rather like the oral tradition, as something that exists in the body of the storyteller and is transmitted to the audience who listens: so they unfold as they are told. As such they all lack stage directions as well as character development, given that the characters themselves are supposed to represent a position and rely strongly on capturing emotions and conveying themselves through the power of the voice.[7]

Following this, many of his children's plays were first broadcast on the radio and later gathered in the collection *The Coming of the Kings and Other Plays* (1970). Based on retellings of traditional material, mainly folktales, fables and myths and written in verse, the plays develop similar themes to those found in his stories and poems, especially the dualisms of rationality/culture and imagination/science promoted by the poet's own Leavisite education and teaching of English. Hughes's attack on the scientific attitude to life and the rationalistic mode of thinking which dominated schools throughout Britain at the time serves as his argument that modern education, by inhibiting the imagination, was destroying children's creative process.

Hughes however, not only believed in imagination as a tool of creativity, but also as a means for regeneration and healing, for both the artist and the reader. In the case of his children's plays it is the listener, through the charge and effect of the spoken voice, who is exercised by endorsing the role of imagination and through the creative experience itself. It would appear that critics have underestimated the art of storytelling and the importance of myth which can be explored in his children's plays. Moreover, the basic fact that he was writing for a younger audience, whom he believed would be able to connect to his world of fiction, by becoming participants of the act of storytelling, is often overlooked.

The essence of Hughes's credo behind his writing for 'Listening and Writing' and the 'Living Language' series of edited radio plays in the 1960s was the basic idea to 'stimulate and foster the listeners' imaginations, by presenting a wide range of material' as Joan Griffiths, series producer, put it (Griffiths and

Kerrison, 1977: 1). The idea was to make the study of literature accessible to all through the art of storytelling and to release the imaginative power of young listeners in order to help them 'visualise' scenes and characters and empathise with feelings. Before working with Griffiths, Ted Hughes had worked with Moira Doolan to devise his contributions to the 'Listening and Writing' series which later became *Poetry in the Making* (1967). Like the programmes that later constituted *Poetry in the Making* (1967), the programmes for 'Living Language' were published as texts with teacher's notes so that the printed pamphlet could be used as a classroom resource for teachers for pupils aged 9 to 11. In them Hughes highlights the importance of the power of the voice and the art of oral storytelling, given the fact that the plays were created with the purpose of being broadcast and not staged. Music and sound effects were also fundamental pieces used to convey different atmospheres or feelings and illustrate the emotional content of the story. However, despite attempts to make the tales sound mysterious this is one element that tended to fall short, possibly because of the strong didactic focus of the plays themselves. The programmes, which were completely novel at the time, were used as a stimulant and as resource material for teachers designed primarily for schools. The main idea was to foster a positive attitude amongst children to the art of storytelling through the experience of listening, speaking to the child's imagination. As Doolan explains:

> Children can enter into the experience of the story; they can re-live the feeling and thought of the writer, and afterwards they can use it as a point of reference in talk for their own experience, or as a taking-off point for their own imaginative creations.
>
> (1974: 2)

The role of the teacher was, of course, also fundamental in offering school children the right forum to foster imagination and giving encouragement and praise. The idea of 'living the story' went beyond the simple experience of listening itself. The programme was also an attempt to make children want to read by entertaining and learning at the same time. In this sense, both his children's plays and the programmes for 'Living Language' as well as 'The Sacred Earth Drama Trust' should be understood as part of the same ongoing project; one in which he brings the stories he tells 'back to earth'.

An ecocritical reading of *The Coming of the Kings and Other Plays* (1970)[8]

Ted Hughes's children's plays read as a reflection of the poet's own art of storytelling. Conceived as part of his own shamanic quest, they illuminate his effort to find the power of the word, the language that unlocks wisdom and what he understands and problematises as the 'truth'. Hughes was fully aware that children could naturally move between the dimensions of fiction and truth, reality and imagination. Folk and fairy tales serve the poet as a symbolic way of presenting the different means by which human beings cope with the world in which they live. He uses them as tools in his writing for children as a way to connect the soul and the imagination, the inner and outer world, but also as a way to heal the split between the child and the adult who has lost contact with the world of myths that children can relate to so easily.

Through these plays we discover the truth behind the poet's own shamanic quest: to tell stories which can, in both a holistic and didactic sense, have a cathartic effect on the reader, or in this case listener by means of healing, regenerating and guiding them. From an ecocritical view, such concerns can be read not only in the light of Carson's sense of wonder but also as sharing similar concerns with the field of ecopsychology.

Broadly defined, ecopsychology acknowledges our biological connection with nature as being therapeutic, focusing on the relationship between nature, language and human psychology, and implies that psychology has a fundamentally important role to play in responding to current social and ecological challenges. By bringing psychology and ecology together, ecopsychology offers a new framework for telling stories as a way to understand the human/nature relationship by creating a healing context.

Oral storytelling can also help increase an appreciation of ecology and the natural world since by listening to stories it allows the young reader to create their own images in their inner-mind's eye. By opening our senses, it allows us to experience and connect the human and non-human world, as well as our inner and outer physical and psychological worlds. In his children's writing Hughes addresses our connection to the natural world, and uses storytelling as a context for healing the human/nature split, that is, as a way of making us environmentally aware of human's alienation from nature whilst offering us the possibility to reconnect with our roots, in order to regain the sense of harmony.

Storytelling constructs a bridge between the past and present and transforms this negotiation into an act of reconciliation. In the case of Hughes, it can be seen as an essential part of his healing quest, as a meeting point between personal life and public performance, that is, a literary strategy to come to terms with events from his own life (a past marked by tragedy and death) and a way to verbalise his trauma, especially in the case of his children's play 'Orpheus'.

'The Coming of the Kings'

The first play of the collection, 'The Coming of the Kings' was presented in two parts on the BBC 'Listening and Writing' Programme for Schools on 27 November and 4 December 1964, and later staged in London in 1972. The frame story is set as a domestic comedy based on the nativity play of the birth of Christ, with a twist to the traditional biblical story of the Three Wise Men and an underlying criticism of materialistic values that suppress the imaginative spirit.

Hughes's employs satire, in the form of irony and humour, which can be seen here in his somewhat simplistic depiction of the characters. Whilst the plot is familiar to the listeners, the materialistic innkeeper and his squabbling wife, who are crudely depicted as comic figures, remain oblivious to the arrival of Mary and Joseph. The play reads as a clear critique of materialism, which blinds the greedy couple, hopelessly incompetent in their own business, highlighting the disregard of the basic Christmas message as a time of special giving.

The opening scene takes place between the greedy couple and a fortune teller, who tells them, through the words of their own talking donkey – who they keep but cannot understand – that 'the King of Men is going to visit your inn today' (*CPlC*: 6). Although the donkey appears to be a minor character in the play, the role Hughes gives it is fundamental. As in fairy tales, Hughes uses the donkey to represent the figure of the animal helper. However, given that the innkeeper and his wife are clearly money-orientated, and not connected to nature, they are both unable to communicate and understand the donkey, and fail to see the real 'truth' that lies before them. In doing so, the poet shows that people who are too involved in worldly things have lost their own connection with nature, cautioning about the dangers of materialism and promoting a change of attitude in the young reader/listener.

In his review of Max Nicholson's *The Environmental Revolution* (1970), Hughes wrote that 'the story of the mind exiled from Nature is the story of Western Man' (in *WP*: 129). At the dawn of the environmental revolution and the early stage of his healing quest it is precisely that alienation from the natural world that Hughes is drawing on here, whilst being aware of the animistic sensibility that Judeo-Christian doctrine and modern worldviews have censored (Sagar, 2009). Just before, Lynn White's seminal essay 'The Historical Roots of Our Ecologic Crisis' (1967) about religion and the environment had been published. In it White claimed that Christianity, combined with science and technology were behind the ecological crisis and that in order to successfully address our environmental crisis, human beings must first examine their own attitudes towards nature. Both Hughes and White appear to voice similar criticisms towards Western values that have alienated us from nature.

Hughes's first stage of educating ecocritical awareness in his playwriting involves exposing the dualities that have disconnected us from the natural world: in this case human greed for materialism and exploiting the Earth, and a loss of touch between human consciousness and nature, or what E. O. Wilson called 'biophilia', defined as 'the innately emotional affiliation of human beings to other living organisms' (Wilson and Kellert, 1993: 31). This is the same alienation that can be seen in the innkeeper and his wife, both of whom are delighted at the perspective of the royal visit and blinded by their greed, thinking that it will make their inn the most popular and famous in Bethlehem. However, without emotional insight, they mistake a pretentious priest, dressed in expensive robes, a corrupt businessman and a merciless police inspector for the three kings. When Joseph and Mary finally arrive they are turned away, and directed to a nearby stable, where the King of the worlds is born in darkness 'to the coughing of animals' (*CPlC*: 24).

The ending highlights the play's contrast between the materialism of the innkeeper, his wife and the guests gathered at the inn, all of whom are only interested in money and appearance, and the imaginative viewpoint of the minstrel and the three kings, who are all in touch with the world of spirits. These differences are also developed not only through the characters and their actions but in the language of the play as well as the different music that accompanies it throughout. As in many of his other radio plays, and bearing in mind the fact that the majority were written with the underlying intention of being listened to, music and song are an important element used by the poet to reinforce dramatic action, evoke atmosphere and define the different characters and situations that appear, as well as helping to set the scene.[9] As for the way the characters speak, Hughes portrays clear differences, not just between the different social classes depicted but through the materialism or spirituality of the characters themselves: the materialistic characters, such as the innkeeper and his wife together with his guests, all speak with comic and grandiloquent language, whilst the minstrel and the kings speak in a more realistic tone. The figure of the storyteller is used here by Hughes throughout the play as an audible narrative voice; in the beginning through the figure of the fortune teller and in the end through the figure of the minstrel, both of whom share the real secret with the 'audience', aware that power and wealth has worn the emotional bond between human beings and the environment, estranging them from nature. One of the most striking elements of the play is the fact that despite being commissioned by BBC radio, unlike the rest of Hughes's children's plays, it contains more detailed stage instructions, thus 'The Coming of the Kings' lends itself more successfully to performance, which could suggest that Hughes actually wrote the play with the idea of staging it in mind.

'The Tiger's Bones'

The second play that appears in the same collection is 'The Tiger's Bones'.[10] The play is without doubt Hughes's most ecological and didactic intervention in the collection. The plot of the story is simple enough: The Master, an imminent scientist has detected a meteor approaching the Earth and is determined to calculate the exact point at which the meteor will make impact. The play highlights the clear division between good and bad, developing Hughes's ideas on the role of imagination and myth as vital tools to restore the balance between the inner and outer worlds of today's society. It draws on the key ideas Hughes himself developed in his essay 'Myth and Education' (1970), especially the conflict between creativity and rational science. In addition, Hughes uses certain techniques of the Theatre of the Absurd, such as the metaphorical significance of names which serve as a label of a person's character: The Master as the personification of science; Jitterwit, who – as his name indicates – talks in short, jittery sentences or Dully, who the others consider to be dumb but is, in fact, the most practical and the only one capable of resolving the mystery of the meteor. Another characteristic, which could be considered as one of its major flaws, is the fact that the play appears to lack a linear development: the play begins and ends, with a growing intensification of the initial situation, whilst jumping from one situation to another resulting at times in gaps in the plot.

As previously mentioned, one of the fundamental characteristics of his children's plays is the fact that most of them are written as dramatic verse for radio. A result of this is that they often read in their printed version as verse play, or are categorised as poetic prose.[11] Given that his children's plays were primarily intended for radio and school performances, it would appear that Hughes was using them as a platform, to voice his own ideas on the importance of imagination as part of the creative process together with his argument that modern education, by inhibiting the imagination, was destroying children's creative process.

The central character of the 'The Tiger's Bones' is a scientist whom Hughes aptly named Master (symbolising scientific objectivity and the rational) who proves in an irrefutable methodical manner that a meteor is about to crash into the earth and destroy it. He sets out, with the help of Von Gonktop, Mr Jitterwit and Dully, the oaf and their porter to 'find the Truth', that is to witness the moment and be at the exact place where according to his calculations a meteor will land. When Jitterwit asks him if the meteor could possibly miss them, Master replies that if a meteor is going to hit the earth then they will simply have to accept it, for 'The Truth is the Truth' (*CPlC*: 32).

Hughes's narrative of the truth progresses in the play and comes through as both a didactic and ecological intervention, underlined by Master's near blindness to see what was happening due to his dependence on science and his condemnation of what surrounds him. He uses the character of Master to criticise modern society's reliance on science, revealing the negative effects of an abandonment of imagination and a reliance on technology, so his own quest of truth here is to show the underlying conflict between the scientist's methods of close, reasoned observation and the poet's free play of imagination. The play also criticises Western colonialism by depicting the colonial control of nature, connecting capitalistic impulses to negative views of nature. Hughes, enacting the role of the storyteller, appears to be parodying adventure stories and the imperialistic dominating view of writers of the Enlightenment (namely Dafoe's *Robinson Crusoe* or Swift's *Gulliver's Travels*) whilst mocking science and technology in order to expose the truth. Indeed, the colonialist values contained in these adventure stories, such as power relations based on conservative, imperial constructions, often privileged human thought and language over nature, as parodied here by Hughes in the character of Master.

On the other hand, the play also shows Hughes's admiration of Swift and the impact he had as a model for his own work. In many ways the journey of the protagonists is reminiscent of Swift's central character in *Gulliver's Travels*. Like Gulliver, they travel through strange lands, 'over mountains, through cities, over seas' (*CPlC*: 34) until 'at last they came to the jungle of red flowers' (*CPlC*: 34) where they discover the 'half-people' 'covered with brown fur' (*CPlC*: 34). The Master first mistakes them for gorillas, since as he exclaims 'there are no savages left. Progress has cleaned up the whole globe' (*CPlC*: 36). However, it is Dully, who corrects him and points out that in fact these are 'very special savages, sir' (*CPlC*: 36). Apart from the underlying similarities with Swift's Houyhnhnms and the savages, the Master's words also echo those of Beckett's Estragon when he bluntly states that 'People are bloody ignorant apes' (Beckett, 2006: 15). Estragon's misanthropic view of humanity underlines that people are driven to actions through habit and sheer ignorance; whilst in the case of Hughes, his quest for the goddess of nature examines human's alienation from both nature and one's true inner self, becoming part of the dualism that he seeks to reconnect in his children's work. This lack of balance between man and nature can be seen when the Guardian Spirit – who narrates the story – explains the way in which the Master insists on helping the savages evolve, first by clearing the forests to sew crops and second by killing all the wild creatures. The play continues detailing how after destroying all the creatures, including the insects that might infect the land Master will poison the land in order to make room for the crops to grow.

This grim depiction of deforestation, together with the social evils of colonialism, has ecological consequences not only on the savages but indeed on the rest of the characters as well. Certainly Hughes's own quest for the truth can be read here as anticipating one of the most urgent issues of our age: that of ecological destruction and pollution. In his play Hughes was cautioning us against the dangers of our own environmental crisis and was already aware of the urgent need to find a more sustainable way for human beings to coexist in Western modern day society. By suggesting the need to heal the split between nature and humanity, Hughes puts forth similar concerns as those voiced by ecopsychologists and uses literature as a tool to educate, warning that humanity is destroying a planet that is no longer sustainable. In the play, echoing humanity's alienation from nature and blinded by his rigid programme of 'progress' and his obsessive quest, Master, by improving the savages, actually enslaves them, bringing on suffering, hunger and leading them to their own quasi-extermination.

When all that remains is a landscape of black ashes, it is Dully, once again, who sees the real truth and condemns his acts. Indeed, Master's scheme to improve humanity, workable in theory, fails in practice, due to his estrangement from nature and his over-reliance on science together with his blindness to see the truth. Dully's speech is reminiscent of Lucky's famous nearly three-page long monologue in *Waiting for Godot* mocking the language of philosophy and science, where the characters also face the chaos of a world that science and logic have long abandoned. However, whilst the core of Lucky's speech refers to the Enlightenment, to the nature of God, the destruction of our world and the degeneration of our species, his pessimism contrasts with Hughes's satirical twist in the play, executed with true Swiftian irony. This can be seen when Dully saves them from the meteor, by unscrewing the telescope and showing Jitterwit the spot of fungus growing on the lens that they had mistaken all this time for a meteor, ridiculing the Master's astronomical observations and scientific calculations. In criticising the Master's ironic blindness Hughes is, of course, condemning humanity's determination and blindness in failing to see reality as it is. In the play they return to civilisation promising not to make the same mistakes again. However, in his attempt to spread the word of truth, the Master causes more desolation and calamity, leading to the destruction of civilisation. The play ends with the Master meeting the sage who lives at the top of the mountain who challenges him to a test of mental strength: to assemble the bones of a tiger. In Hughes's tragic-comic ending the Master is, of course, destroyed by his own intellectual arrogance. By giving life to the bones of the tiger, the life he creates puts an end to his own. Moreover, as Yvonne Reddick has correctly suggested, the resurrection of the tiger at the end of the play could also have been echoing Hughes's growing concern with endangered species (2017: 181).

Notwithstanding 'The Tiger's Bones' also has its flaws. There are two weaker points that are particularly significant, given that they underlie nearly all of Hughes's children's plays: the poet's lack of insight in developing characters and his excessive use of narrative voice. The first of these apparent weaker points can be explained by remembering that Hughes used his characters very much as symbols and archetypes to voice his opinion. There is a clear didactic connotation underlying the characters, who serve throughout the plot as stereotypes for different behaviours. Hughes uses his characters in a similar way as in folktales, that is, largely as universal symbols: the good, the innocent, the evil or as archetypes to enact role positions, such as the hero, the trickster, the wise man or the damsel in distress, rather like symbolic motifs in poetry. Likewise, animals, myth, folklore and archetypes, serve as source of symbols and objective correlatives of Hughes's own feelings and ideas. Bearing in mind that the plays were not written with the intention of being performed on stage, but rather to be read on the radio, Hughes's lack of character development does not come across as being a serious problem, especially if we consider that their main underlying aim was to stimulate the imagination of the listener – a children's audience – so that dialogue, together with Hughes's use of verbal humour, become the most important aspect.

The same, however, cannot be said for Hughes's excessive use of narrator, who anticipates events and at times interrupts the dialogue of the characters, impeding the play's natural flow. This can be seen in 'The Tiger's Bones', where Dully's somewhat disjointed monologues are essentially narrative, and appear at times, to overstate the message. A clear example of this occurs at the end of the play, when Dully exclaims 'So now I'm going home, none the wiser. I'll have to think of a moral to this story' (*CPIC*: 51), and just in case the audience does not get the moral either, it is the Guardian Spirit, an over-intrusive narrator one might say, who provides it by closing the story with Dully returning to his father's farm and restoring the story's circular frame in harmony with the image of initial pastoral peace.

It is also important to point out that Hughes's plays can, at times, be considered as being almost too didactic, in the sense that by relying on essentialising stereotypes they are intentionally one-sided. Rather than attempting to unify dualisms, at this early stage of his healing quest Hughes simply limits himself to exposing dualities and giving his view on opposites such as science/education, nature/culture, etc.[12] However, despite its minor flaws, 'The Tiger's Bones' reads as a brilliant satire on the excess of rationalism and science, in which Beckett's absurd humour, combined together with Swift's satire, is used successfully as a creative tool by Hughes.[13]

'Beauty and the Beast'

'Beauty and the Beast' was broadcast on 2 December 1965 and first staged in London in 1971. Hughes varies the traditional tale by incorporating elements of fable, folk and fairy tale. The story begins with a floor sweeper, who acts as the narrator, and represents the lower classes, telling how a millionaire's daughter, Floreat, powerless and sick, appears to be 'bewitched' by a monster that stalks her at night and suffers from some inexplicable ailment. Once again Hughes mocks scientific objectivity, here through the figure of the doctor, who tries to cure his daughter with various scientific remedies, and also criticises materialism: for all his money the father cannot heal his daughter.

In his attack on scientific rationalism, and the conflict between the scientist's method of close reasoned observation and the poet's free play of imagination, Hughes claimed that by suppressing instincts and through its alienation from nature, modern civilisation was suffering from a self-inflicted malady, one caused by its own excessive cerebration, resulting, in the words of Annie Schofield, in a 'chronically sick society' (1983: 33). Hughes is drawing here on the healing capacity of the human psyche following the Jungian school of Depth Psychology,[14] in line with Jung's belief that psychological distress is a result of an imbalance within the individual.

In his version of the fairy tale Hughes defends the need to remain in touch with nature and uses his storytelling as a powerful tool for healing the human psyche. Floreat's father brings in the best doctors to cure his daughter but all of them fail to see that all she needs is love. Meanwhile the 'beast' appears in the form of a huge, ugly black monster that comes to Floreat's room every night. One night, after hiding in his daughter's room, Beauty's father shoots the monster in the arm and believes it to have died in the forest. The next day a bear-keeper comes to their mansion with an awkward-looking dancing bear, called Bonzo, with a bandaged arm and a pet canary. Despite Floreat's father's initial suspicions, the clever bear manages to trick the father together with his son, not only by freeing himself from his imprisonment, in the form of popular street entertainment, but by seducing Floreat and running off with her in his arms into the forest. For three nights he keeps her in a cave, feeding her with wild fruits. Not quite sure what to make of the situation, Floreat, strongly reminiscent of Little Red Riding Hood and her encounter in bed with the disguised wolf, acknowledges that he seems a gentle creature and begins to befriend the bear, overlooking his ugliness and wild nature.

At this point Floreat's father, together with the hunters, enter the scene ready to rescue her by killing the bear. However, Hughes's play offers a twist when Floreat, appearing to suffer from a form of Stockholm syndrome, stops them from shooting the bear just in time after realising that she is cured and confesses her true feelings of love towards the beast: 'Look at me. I'm better. I'm happy. I'm cured' (*CPIC*: 70). As in the original version of *Beauty and the Beast*, the psychological bonding that Floreat experiences towards the bear is similar to that of Belle who also shows sympathy towards the monster and falls in love. However, in Hughes's tale there is a much more implicit ecological moral. Only when there is communication between Floreat and the monster is she cured and at the same time does she break the spell that the young man, in the form of the beast, is under. As revealed, it is when Floreat is removed from her domestic environment and is reconnected with the natural world that she is cured and so is the rest of the community.

In the poet's tale Floreat's healing can be seen as a consequence of her own reconnecting with nature. Advocating caring for the earth, as ecopsychology encourages, also entails using the tools of depth psychology. As depth psychologists with an ecopsychological emphasis hold (Aizenstat, 1995 and 2003; Kahn, 1999; Roszak *et al.*, 1995), our physiological illnesses are connected to our damaged relationship to nature and human alienation from the natural world contributes to our physical suffering. In the play, Floreat's true confession of love for the beast not only saves him from death, but, of course, restores the beast's humanity and acts as a symbolic expression of healing. However, it is not compassion, as in the fairy tale, that turns the bear back into a human being, but Floreat's true act of passion. In many ways, as in Grimm's version of 'The Frog Prince',[15] the play tells of a transformation brought about by an act of love. Transformation is, of course, at the heart of all magical practices, and appears in many folk and fairy tales.

Following this, the play has a happy ending with the father giving his daughter in marriage to the hero, in this case the bear, transformed into a handsome young man. The marriage of Beauty to the beast is, in the words of Bettelheim, 'a symbolic expression of the healing of the pernicious break between the animal and the higher aspects of man' (1977: 308–309). Indeed, in comparison with his previous plays, 'Beauty and the Beast' proves to be a far more eloquent and effective healing tale, thanks to the progression of Hughes's playwriting technique, which can be seen here in the way in which he uses storytelling to deal with the split between human/nature, through the figures of Floreat and the bear. Whilst Floreat suffers nightmares and her millionaire father believes that she is bewitched, it is in fact the bear who is the monster in the night and has been put under a spell. The shadow that haunts Floreat is the shadow of love, the same love that will cure her when it is released. It is precisely the recognition of this, when she tells the bear she loves him, that allows both Floreat and the bear to be freed. This act of transformation, the metamorphosis of the bear/prince allows Hughes to create connections across the human/animal boundaries. Thus, by transforming the beast from animal/human and healing Floreat, he successfully unifies the duality of human/nature and is able to reconnect the human/non-human, thanks to an act of true love, allowing for redemption to finally be obtained.

'Sean, the Fool, the Devil and the Cats'

'Sean, the Fool, the Devil and the Cats' was first broadcast on 2 May 1968 and later staged in London in 1971. Like Hughes's other children's plays, it combines elements of folklore, myth and fairy tale. The protagonist, Sean, takes on the role of the hero who undertakes a personal quest, leaving home and accepting the call to adventure. The fact that Sean leaves both his home and his possessive mother and sets off to seek his fortune in the outside world can be seen as representing the transition of childhood to adulthood. Sean soon finds himself in a riddle, walking at the roadside and wondering where the different roads go: 'Roads are a maze with no ending' (*CPlC*: 76).

Hughes uses folklore here to show the reader the different paths in life and the consequences of choosing one road over another,[16] so the play can be read as an allegory for life. The imaginary roads represent the different decisions in life and the play tells the story of what happens when the rational is abandoned and instinct is followed instead. The element of suspense comes in the play as the listener/reader wonders which path Sean will take. He is faced with four different roads to follow: the first one offers him happiness, the second the prospect of adventure, the third, recalling the voice of his mother, beckons him home and the fourth promises fortune and a secure future. Sean's determination is given a dramatic tone by Hughes as he disregards all the voices, including his mother's, which lure him to a different path, and chooses instead a fifth, that of the dark wood. The imagery of darkness reveals that it will be the most threatening, as do the voices that plead with Sean to avoid the path of the unknown. At the crossroads Sean meets the fool, who acts here as the antagonist, digging in the darkness of a ditch, looking for an enormous treasure. After exchanging his sandwich for the fool's luck Sean follows his instinct by choosing the path that leads to the devil's house. Along the path Sean meets a strange huge black cat, which leads him to the devil's house, somewhat loosely depicted in the play as 'Round, stone, a strange house' (*CPlC*: 80). As in traditional folk literature the hero is tested to prove his identity, and is aided by a magical agent or helper. Hughes uses once again the figure of the animal helper in order to guide the hero, in this case Sean, in his quest. As in 'Beauty and the Beast' a spell has enacted a transformation from human/animal. The hero, by accepting the call, undertakes a spiritual journey, like the shaman, and is faced with the task of resolving the riddle and winning the quest in order to succeed.

Given its limited set of recurrent actions and archetype figures, the development of the rest of the play is somewhat predictable: Sean finds out that the devil has cast a spell and transformed Miss Poppacopolis, a rich man's beautiful daughter, into a cat, and has to find a cure to save her. The devil spends the next three pages explaining, rather exhaustively, how to lift the curse: the girl's father, who owns ten factories, must share all of his profits with the workers, who earn low wages, and then allow each of them to hit him with a stick or umbrella. For the last step in the cure he must find somebody willing to be turned into a dog to chase his daughter to the top of a tree, where, when she hears the barking, she 'Will come to her senses and be cured' (*CPlC*: 84).

It would appear that one of the reasons that could account for the somewhat long-winded and over-descriptive explanation is the fact that through his narrative, the poet is trying to demonstrate to his audience in a very visual way, the power of oral storytelling as a way to teach lessons about the world, and heal the psyche, so that the 'internal processes are externalised and become comprehensible as represented by the figures of the story and its events' (Bettelheim, 1977: 25). Throughout Hughes's storytelling there is an underlying necessity for the reader to believe the tale and integrate it into their own experience, and so, by confronting the unknown, the reader, like the hero in folk and fairy tales, can imaginatively win the quest and be healed.

The character of the fool as the voice of truth recalls the Shakespearean fool, acting here as the narrator of the story and tells how Sean, enacting the quest of the typical hero, completes each of the steps and finally, as in traditional folklore, and following 'Beauty and the Beast', gets to marry the daughter. However, the relationship between the fool and Sean is not totally clear in the tale and lacks any in-depth development. The fool, in his jealousy after listening to Sean, also decides to try his luck and outdo the hero, and follows the same path that Sean took to the devil's house. However, this is a cautionary tale about eating the wrong person's stew, and the fool's luck does not obviously have the same outcome. As in traditional folk tales, only the true hero can succeed, and instead the fool's quest ends in comic failure, for in his haste, he forgets to give stew to the cat, which bites his ear and starches out his eyes. Psychologically, the fool's blindness could symbolise his lack of, and need for, consciousness. It could also represent in the antagonist's journey of self-knowledge, the abandoning of the quest and, consequently, a loss of faith.

Throughout the play, Hughes effectively allows the protagonist to follow his imagination, to explore the tentative path he chooses, and in doing so he succeeds in his own quest. Hughes's children's plays can be read as stories that spark the imagination and provide children with tales that form part of the creative force behind Hughes's own quest of truth. By using folk and fairy tales as part of his shamanic quest, the poet gives a voice to nature, breaking what Christopher Manes calls 'the silence of nature' (1996: 16) and is able to successfully explore the underlying tensions between human nature and the natural world and effectively reconcile both, given that the treatment of the inner subject also requires attention to the outer world. Hughes uses the same technique as with Floreat in 'Beauty and the Beast', so that the deconstruction or 'displacing' of the subject involves re-placing the subject back in the world, reconnecting the nature of the soul, and in the case of Sean, finding the right path to follow.

For Hughes, finding the right words to tell the story was of vital importance. The voice of the storyteller had to be heard by the reader to be brought to life, and transmit experience. Drawing on the importance of emotion, feeling and imagination, the poet himself explains this process of thinking, 'by which we break into that inner life and capture answers' (*PM*: 57) and its link with creative writing, cautioning us that 'if we do not somehow learn it, then our minds lie in us like the fish in the pond of a man who cannot fish' (*PM*: 58). Of course, for the poet, the child was more open and closer to imagination than the adult who had forgotten how to *fish* inside that inner world. Anticipating Wilson's biophilia, Hughes believed that children also had a special affinity for nature, so that early experiences with the natural world can be seen as being positively linked with the development of imagination and the sense of wonder (Carson, 1965; Cobb, 1977). This way of understanding, as Hughes recognised, also serves as a motivation for further learning. Thus the art of storytelling and the importance of voice that lie behind his children's plays are strongly linked to the world of imagination and can be seen as a means of articulating experiences and creating the myths we need. If as Ralph Metzner has noted, 'Western modern child-rearing practices effectively stifle any innate ecological sensibility the child may have' (1995: 63), then storytelling can be seen as an important tool to foster ecological consciousness. It is worth remembering that traditionally, as David Abram has noted, in certain cultures the oral storyteller was believed to hold magical powers and even the stories themselves were believed to be enchanted. In this sense, Hughes sees himself as an oral storyteller, a poetic writer with a purpose, and uses the radio as a means to express his ideas and spread through the power of myth, the aspects of 'truth', as part of his own shamanic healing quest.

'Orpheus': Hughes's own shamanic journey[17]

'Orpheus' is perhaps the most successful, and indeed the most redemptive, of all of Hughes's children's plays. It was first broadcast on 29 January 1971, and later published in the American edition of *The Coming of the Kings*, titled *The Tiger's Bones*, in 1974. However, it did not appear in England until the posthumous publication of his *Collected Plays for Children*, in 2001.

Critics have often overlooked the importance of Hughes's early children's plays and despite the fact that 'Orpheus' reads as a personal account of Hughes's emotional struggle following Plath's death, it has received very little critical attention. In Sagar's *The Laughter of Foxes* (2006), the 'Orpheus story' is the only children's work of Hughes that gets any mention. As Sagar points out, Hughes purposely avoided the story in his writing for adults for over two decades 'even conspicuously omitting it from his *Tales from Ovid*' (2006: 85). Although Sagar fails to explore here the fundamental question of why Hughes chose to write the story for children and not adults, the answer lies in a letter the poet wrote to Sagar in 1998, just months before his own death. In it he confesses that it was the first story that occurred to him after Plath's death, but that he had rejected the story on the grounds of it being too obvious to use: 'I was too conscious of that obviousness. I saw my little note about it, the other day.

The shock twist was that Pluto answered: No, of course you can't have her back. She's dead, you idiot' (Sagar, 2012: 275).

Hughes's personal confession to Sagar is both moving and revealing, especially bearing in mind that Hughes wrote this at a time when he was aware that he was near the end of his life. This would also suggest that the poet's writing for children belongs to a much more personal project, that of his own healing quest. The play reads as an account of Hughes's own imaginative journey to recover his dead wife and as a description of the catharsis of the healing process itself.

Despite the fact that Hughes's initial idea was to write a version of 'Orpheus' after Sylvia Plath's death, it seems fitting that he should come to terms with actually doing so after finishing *Crow* (1970) and the year after the traumatic period in which his mother died, as well as Assia Wevill and Shura.[18] It reads as a personal account of Hughes's emotional struggle in the years following Sylvia Plath's death, and his coming to terms with his own personal torture and confession of how her death had left his creative life stagnant. It must be remembered that after Plath's suicide and until the publications of *Wodwo* (1967) and *Crow* (1970), Hughes did not publish any literature for adults whatsoever. However, it was during this period that some of his most important children's works were created.

So, although 'Orpheus' should not be read as a play for adults – Hughes purposely avoided the Orpheus story in his adult work for over two decades until the posthumous publication of *Alcestis* (1999) – it can be read as a means of his coming to terms with the reality of his own traumatic personal situation. Hughes himself argued that myths contained healing characteristics that reunite our inner world with our conscious life. Given the underlying cathartic properties of his play, it reads as an essential part of the poet's healing quest to recover his own creativity. Orpheus is, like Hughes's own struggle, a shamanic journey into the underworld where Hughes takes on the role of Orpheus, and Plath that of Eurydice. It also reads as a journey to recover the truth and one where the shaman reconnects with the sense of the 'sacred' within nature. As Barry Powell states,

> The myth of Orpheus [is] more than a simple folktale, the story of a man who went to the other world and returned, because we can view Orpheus as a shaman in his own right, returning from the underworld to teach the truth about human nature and destiny.
>
> (2004: 306)

Hughes's version of Orpheus draws strongly on the most familiar versions of the myth, those of Virgil and Ovid. Like Hughes's version these are written in the form of poetic prose and tell a similar tale of doomed love and guilt, although the endings are different. The tragic tale tells how the newlywed lovers are torn away from each other by death. The myth of Orpheus has also been widely considered as an archetypal myth that symbolises the healing powers of music (Marchenkov, 2009). The beginning of the play celebrates the cycles of nature and centres on Orpheus's magical music that makes the trees and stones dance. His music has the ability to charm all living things and reaches 'to the bears in the forest', 'the deer in the wrinkles of the hills' and 'the salmon in the pools below the falls' (*CPlC*: 95). Not only is the role of music symbolically important in 'Orpheus', through the fusion of words with music to depict the different affective emotional states of the character, but it acts as a fundamental link between that of Hughes's written word and a way of arousing affection in the listening experience of the audience.

In classical mythology, Orpheus's music could charm all living creatures in nature, even trees and rivers, or rocks. Orpheus was after all, one of very few Greek heroes who returned from the underworld and whose music even held power over Hades. In Hughes's play the music transforms itself with the character: it begins as energetic pop music up to the death of Eurydice, and then jangles into discord and becomes erratic, changing to electronic music. In the myth Eurydice is fatally bitten by a snake. In Hughes's version Orpheus's hand suddenly becomes numb and he can no longer play music. At the same moment he discovers that Eurydice has died from a snake bite. The numbness of Orpheus's hand would symbolically appear to represent Hughes's own 'writer's block' and the creative numbness that he experienced after Plath's death. Orpheus refuses to accept that he will never see his beloved wife and descends to the underworld to persuade Hades to win her back. After his descent to the underworld 'through the bowels of the earth' (*CPlC*: 101) he lands at the feet of Pluto, king of the kingdom of the dead, and asks him for the return of his wife. Pluto reminds him that his wife was the price of his music. Pluto's own wife, Persephone, is doomed to spend six months on earth where she 'is a flower-face, she laughs and sings; everybody adores her' (*CPlC*: 102) and six months in the underworld where she has 'the face of a maggot' (*CPlC*: 102). Orpheus's music causes Persephone to flower and for the first time Pluto is able to see her face open. Following this, he offers to release her under the condition of 'A wife for a wife' (*CPlC*: 103). However, Pluto is unable to keep his promise and Eurydice's body crumbles to dust, like Lot's unnamed wife in the biblical exodus scene out of Sodom and Gomorrah, as she turns to witness the destruction of her home. Notwithstanding, in Hughes's version he does allow Orpheus to return to earth with her soul.

In Virgil's version of the myth (see Book 4 in *The Georgics*), Orpheus is allowed to take Eurydice back on the condition that he walks in front of her and does not look back until they both reach the upper world. Before he reaches the upper world, Orpheus glances back to confirm Eurydice is following him and at that moment she vanishes and Orpheus loses her forever. Orpheus's music had the power to save Eurydice, but fails to do so. In Hughes's version of the myth, Orpheus fails to recover his dead wife but, instead recuperates her power of speech. That Hughes recuperates Eurydice's voice can be read symbolically as the poet's coming to terms with Plath's death. In Hughes's version, when Orpheus looks back he only confirms what he already knows: that Eurydice, like the poet's own beloved and creative partner, Plath, is no longer there. Like Orpheus he can no longer see her or touch her, but her voice and the power of the words that she wrote remain.[19]

Hughes's ending differs from that of Virgil's and bears more similarities to that of Ovid's in that it does not include the grisly details of Orpheus's death,[20] instead the musician is left to wander and doomed to play the music of 'love going/And love lost forever/The music of birth and of death' (*CPIC*: 105). The healing power of music implicit in 'Orpheus' is of course closely linked to the power of rebirth, and in the case of Hughes, it can be read symbolically as the quest for the poet to regain his own voice whilst coming to terms with his own tragic loss.

Following this, 'Orpheus' should be read as Hughes's own personal journey and his quest for truth. Hughes, like Orpheus, is attached to the past, in his coming to terms and refusal to let go following Plath's tragic death. It was of course this same sense of grief which led to the poet's own loss of inspiration and poetic stagnation following his torment of creative inability to conceive any vision of the future. By rewriting the myth in the form of a children's play it allowed him to explore the emotions and feelings of Orpheus by focusing on his own emotional journey without the restraints of writing for an adult audience.

For Hughes, not only is children's literature the perfect vehicle for enacting such a myth, but, in line with ecopsychology, it offers the ideal context in which to achieve a balance between the dualities of the inner and outer, human and animal worlds. Hughes himself recognised that 'Children are most sensitive to the inner world, because they are the least conditioned by scientific objectivity to life' (*WP*: 149). As for why the poet chose to rewrite the myth of Orpheus, it would seem that Hughes's own emotional connection with the myth allowed him to express so clearly the pain and torment of Orpheus.

Hughes himself argued that myths contained cathartic properties that can enact as a framework for the energies of our inner world to reconcile with our outer world. He saw myths as visionary narratives that stood the test of time, bringing order and balance to the incoherent opposing forces of the inner and outer worlds. When asked whether he thought a myth was 'something that we can invent?' (Hughes, 1970: 68), he replied that we only invent the myths we need to heal ourselves and that 'You think of one myth rather than another because that myth is the one that belongs to you at that moment' (Hughes, 1970: 68).

So for Hughes, who uses the ancient and universal material of mythology as a means of dealing with his most personal problems, the myth of Orpheus seemed to offer the greatest possibility of healing to the poet. Following this, Orpheus depicts the poet's own shamanic ritual of healing, and his own personal redemptive quest to resolve the pain and possible sense of guilt after Sylvia Plath's suicide, enlarged by Assia Wevill and Shura's deaths. As the poet shows us through his own quest, the process of articulating truths, of taking possession of the reality, no matter how painful that may be, is both restorative and healing. His writing for children is, after all, one that pursues healing truths. In this light,

Orpheus should be understood as a fundamental part of Hughes's own healing quest, and was written as both a cathartic and regenerative exercise, enacting Hughes's own faith in the power of poetic imagination and as a way to heal his own grief and personal situation after Plath's tragic death.

Indeed, another reason why 'Orpheus' reads as such an effective redemptive myth is precisely because of the dismemberment that Orpheus suffers in his journey to the underworld, depicted in the changing music, no longer 'the music of happiness' but instead 'the music of birth and death' (105). In his essay 'Regenerations' (*WP*) Hughes explains how this transformation works and how the shaman must suffer dismemberment as a necessary part of the shaman's initiation ritual. Following this, the dismemberment of Orpheus symbolises the death of the poet's ego that must occur before he can find his own true voice, reconnected, humbled, towards nature. In the weeks following Plath's death Hughes wrote two animal poems in which he attempted to find that voice. Unlike his previous animal poems, both portrayed animals as victims: 'The Song of the Rat' and 'The Howling of Wolves', which was inspired by the wolves that could be heard from Plath's flat, which Hughes moved to after her death, near Regents Park Zoo. It describes how the caged wolf 'goes to and fro, trailing its haunches and whimpering horribly' (*Wodwo*, 1967: 178). It can also be read as Hughes's first poetic attempt to come to terms with losing a wife to suicide, but the poet's shamanic quest and Hughes's own personal myth are, of course, unattainable in these bleak poems, as Chen Hong has noted 'The shamanic spirit stops working here where the animal becomes a pitied victim rather than a totem guide' (2011: 49).

It is, of course, significant that Hughes avoided the story in his adult writing until his last work, the play *Alcestis* (1999). Given that Hughes was aware at the time of completing the final draft that his cancer was terminal, *Alcestis* can be considered, like *Birthday Letters* (1998) as one of his most confessional works. The last lines express the poet's hope whilst wrestling with death himself:

> Nothing is certain.
> What had seemed inevitable
> Comes to nothing....
> Let this give man hope.
> (*Alcestis*, 1999: 83)

The emphatic monosyllables bring the play to an end; however, despite distilling a rather muted but positive ending, the play itself is by no means reassuring, and depicts a world full of violence and confusion. Unlike *Alcestis*, 'Orpheus' is Hughes's enactment of his own most cathartic and personal myth, that of recovering the poet himself. As Orpheus, Hughes must undergo the shaman's journey to the underworld. According to Hughes the shaman's journey took place in order 'to guide some soul to the underworld, or bring back a sick man's lost soul' (*WP*: 57). As he shows us through his own quest and the cathartic journey of Orpheus to the underworld and back, the process of articulating truths, of taking possession of the reality, no matter how painful they may be, is both restorative and healing.

Like Floreat, in Hughes's adaptation of 'Beauty and the Beast', the metamorphosis that Orpheus suffers can also be read in line with the healing function of storytelling. Metamorphosis is central to all fairy tales and myths, where characters are able to cross borders and undergo changes. As Bettelheim (1977) revealed, the therapeutic uses of enchantment in fairy tales also have the power to change both the reader and listener. So the healing function of storytelling and the idea of personal transformation would appear to go hand in hand. Benjamin stresses how storytelling originates in suffering, death, or feelings of displacement, whilst being at the same time the solution to overcoming such feelings. Donald Ward has also explored the therapeutic power of storytelling and explains just how important narrative can be in conditions of deep crisis and personal disorder, so that 'when these stories are finally told, they can provide the vehicle for needed social transformations' (1990: 41).

So it would appear that Hughes's personal dramatisation of the legend of Orpheus and his own enactment of the shamanic quest journey take place more effectively in his children's work. Hughes's interpretation of the myth has a cathartic holistic effect restoring the balance and harmony necessary in his own personal sphere as well as that of his creativity as a writer, allowing him to continue his shamanic healing quest in both the personal and public realms.

Representing nature and listening to the earth

Throughout his children's plays Hughes portrays connections between the human/non-human worlds and often represents the conflicts underlying different dualisms such as that of nature and culture: 'The Coming of the Kings' is a fable about the dangers of materialism; in the same way 'Sean, the Fool, the Devil and the Cats' not only highlights the importance of imagination, but also shows the dangers of greed and materialism depicted in Mr Poppacopolis and the exploitation of factory workers. In the same vein, 'The Tiger's Bones' enacts the reliance on science and the abandonment of imagination, and reveals our responsibility for what happens to nature and to the other. Furthermore, it criticises ecological and social injustices through the character of the Master and shows the effects of technology and science on community and society.

His children's plays portray the interdependence between nature and culture, and highlight the importance of being attuned to nature and 'listening to the earth'. Many of his collected plays reflect ecological debates and transform them into what could be called 'ecodrama', given the way Hughes engages in environmental issues through the radio performance itself, that is, through the interaction that takes place between the oral storyteller, and audience in his children's plays. Despite the fact that the plays were written for the radio and not intended to be performed on stage, Hughes appears to embrace what Una Chaudhuri in the 1990s described as 'a possible ecological theater' (1994: 24), one in which 'the theater can become the site of a much-needed ecological consciousness' (1994: 28).

As a storyteller, Hughes uses drama as a way to explore the underlying tensions between inner and outer nature, that is, as a way to understand human nature and our relationship to the natural world. Moreover, his 'ecodrama' is reinforced by the fact that he uses the framework of oral storytelling to explore ecological issues, by considering themes such as the dangers of an excessive reliance on technology and the consequent abandonment of imagination; the material impact of capitalism; the importance of community; human means of resources and production; our relationship to the natural world; or looking at how industrial society has caused today's ecological disaster.

His children's drama can also help to raise awareness about ecological issues, and challenge us to think about who we are and our relationship and place in the ecosystem of life. His heroes are simple characters like Dully, Floreat or Sean, who all share one common feature: each and every one of them is in harmony with nature. From an ecocritical viewpoint, Hughes's representation of characters can help transform our conception and attitudes towards ecological reality. Through myth and oral storytelling's potential reconnecting to our inner self, his ecodrama can shape and change our political and social perceptions about environment and our experience and relation with nature, raising consciousness for change, so that as May observes: 'when we leave the theatre, things around us are more alive, we listen better, and we have a deeper sense of our own ecological identity' (2005: 95).

Hughes also clearly develops the sense of environmental ethics in exploring the nature/culture binary in his children's plays. Despite the obvious fact that drama tends to be largely human-centred, Hughes successfully succeeds in bringing the 'other than human' world into play by developing a sense of interdependence with the land and natural world and fosters an ethic of responsibility for the natural world. In his plays he often draws on the figure of the animal helper or donor. In fairy tales, the donor is an animal character that tests the hero and provides assistance, guiding the hero in his search. As Donald Haase has observed, the origin of magical animal helpers in folklore is rooted in ancient rituals and plays a similar role to that of the shaman: 'The allotted function of such a person in a community was to undertake quests in the otherworld by way of helping members of the community to solve their problems, thus focusing on the issue of magical help' (2008: 596).

In 'The Coming of the Kings' it is the donkey, together with the ox, cock and dog that inform the fortune teller that the three kings are coming. Hughes draws on verbal communication and animal sounds (the donkey 'brays', whilst the ox 'lows', the cock 'crows' and the dog 'barks') which are incomprehensible to the owners of the inn and need to be interpreted by the narrator of the tale. In 'Sean, the Fool, the Devil and the Cats' it is the cat, another domestic creature, who takes on the role of the donor, and helps Sean reach the Devil's house. As in traditional fairy tales, where human beings can also be represented as animals or enchanted beasts, in 'Beauty and the Beast' the prince has been transformed into a bear; in 'Sean, the Fool, the Devil and the Cats' it is Mr Poppacopolis's daughter who has been turned into the devil's cat. Indeed, shapeshifting, as Marina Warner has pointed out (1994: xix), is one of the defining features of fairy tales and occurs in ways that contradict the laws of nature.

Hughes's poetic imagination draws strongly on animism throughout his children's work as a creative force which he uses to connect the human/animal world. In animistic knowledge there is no distinction between living and non-living things, since not only creatures, but mountains, rivers, rocks and all other inanimate objects are believed to have souls. In 'Orpheus', even the stones and trees become animated objects. According to Bettelheim, who draws on Jean Piaget's idea of child animism, if young children believe that all objects, both animal and material, have a life of their own, it seems only natural that non-human characters can talk and cross over between the boundaries of the human/animal world. Following this, it is entirely plausible when inanimate objects begin to talk 'since there is not a sharp line drawn between living and dead things, the latter, too, can come to life' (Bettelheim, 1977: 47).

Like Bettelheim, Hughes was strongly aware that children are more attuned to the natural world around them, whilst adults have become 'estranged from the ways in which young people experience the world' (Bettelheim, 1977: 47). So for the poet, myths and fairy tales were an ideal framework within which to explore the rifts between the adult and child (reflected by the different ways of listening to the plays), the human/animal world and above all, to connect and heal our inner and outer lives. Moreover, by listening to the plays, and by becoming participants, children imaginatively take on the roles as characters, and are able in this way to connect between the inner and outer worlds of fiction so that the stories *come alive*. A clear example of this lies in 'Orpheus' where Hughes, as the shaman and poet, connects these worlds. In 'Myth and Education' Hughes explains the process by which the inner world can only be reconciled in the presence of the outer world and describes how imagination is the faculty 'which embraces both outer and inner worlds in a creative spirit' (*WP*: 150–151).

Hughes's ecodrama is closely linked to Carson's sense of fostering environmental awareness and transforming our perceptions and attitudes towards ecological reality by inspiring the reader (or rather listener in this case) to rethink their relationship with nature. Indeed, the ecological link that the infant develops with the natural world is one that is created in childhood and one that is critical to our development with how we see nature.

As Carson and as ecopsychology would recognise, Hughes's ecodrama offers a fresh framework for schools by exploring through his plays our experience and relation with nature, illustrating the bonds and exploring the underlying connections between people and place, culture and nature and the human/non-human world. Hughes's playwriting recovers and reconnects the reader (both the adult and child) to Carson's sense of wonder, through the experience of oral storytelling, using imagination as a tool to foster environmental awareness. Following this, Hughes may have also been suggesting that the adult listener could be healed by, and as, the child reader. His ecodrama was not just directed at fostering environmental awareness in future generations, but also conceived as a way of making adults more receptive and open to their inner self, for letting down the barriers and reconnecting their imaginative and creative energies, in order to embrace both the natural world and the lost child within.

Notes

1 Theresa May first coined the term in 'Kneading Marie Clements' *Burning Vision*', *Canadian Theatre Review*, 144 (Fall 2010), pp. 5–12, where she defined ecodramaturgy as 'play-making (both script development and production) that puts ecological reciprocity and community at the center of its theatrical and thematic intent' (7).

2 My thanks to Zoe Bennett, Pauline Mayne's daughter, and to Steve Ely for their email communications over a number of months in 2011 and their help and insightful comments about Ted Hughes's school days, especially as a sixth-former at Mexborough.

3 Tradition held that the staff performed a revue for the sixth form and the sixth form returned the compliment. Both revues comprised a series of absurd, pseudo-Monty Python sketches.

4 It was also at Mexborough that Hughes initiated himself as a literary critic when, in 1948, he wrote a review for *The Don and Dearne* on the School Dramatic Society's *The Importance of Being Earnest.*

5 Hughes and Plath often frequented the theatre in London as can be seen from their correspondence which mentions seeing John Arden's *Serjeant Musgrave's Dance* (1959), Harold Pinter's *The Caretaker* (1960), W. S. Merwin's *The Gilded West* (1961) as well as other playwrights of 'kitchen sink' drama and social realism, including John Osborne, or Arnold Wesker.

6 Hughes's ongoing quest for a lingua franca had already begun back in 1969 in his earlier collaboration with Peter Brook. In the summer of 1971 Hughes accompanied Peter Brook to the Shiraz-Persepolis Festival in Iran, where he wrote and staged his widely acclaimed *Orghast*, a mythic play written in an 'international' language, a kind of lingua franca that Hughes had invented for Brook's company, the 'International Centre for Theatre Research'. Brook had employed actors with a range of backgrounds and nationalities, believing that this would produce something completely new. The project was to create and perform a play with a new language – a language that united sound, gesture and written word together in a powerful story. Hughes was to write the experiment, in the form of his mythic play.

7 The plays, which came in class sets and were collected in the pamphlets 'Living Language' and 'Listening and Writing', were produced by BBC Radio for Schools during the 1960s and 1970s as a free service for secondary school teachers to use and were ordered by schools. Even when the plays were collected as material for classroom practice in schools, the instructions that came with the teaching pamphlets were for recording the plays, and not for acting them.

8 Hughes's *Collected Plays for Children* (2001) contains four plays originally published in the collection *The Coming of the Kings and Other Plays* (1970): 'The Coming of the Kings', 'The Tiger's Bones', 'Beauty and the Beast' and 'Sean, the Fool, the Devil and the Cats'. 'Orpheus' (1971), previously only ever published in America, and 'The Pig Organ', originally written as a libretto with music by Richard Blackford and premiered at the Roundhouse, London 1980, are also included. 'The Pig Organ' has been excluded from consideration here, given the fact that, unlike the other plays in the collection, it was written with the intention of being staged. This chapter focuses instead on those plays that best illustrate Hughes's healing quest.

9 Hughes comically pairs the businessman with the other guests to sing the well-known children's nursery rhyme 'Three Blind Mice' as a round-song, mocking their supposed intelligence; in contrast when the three kings later appear they are fittingly accompanied by the familiar and traditional Christmas carols to mark the occasion, first 'We Three Kings' and later 'Once in Royal David's City' in order to announce the birth of Christ.

10 The play was first published in *The Coming of the Kings* (1970), in London by Faber and Faber and in New York by Viking, in 1974, and also presented in two parts on 26 November and 3 December 1965 on the BBC 'Listening and Writing' Programme.

11 A clear example of this is *Nessie the Mannerless Monster* (1964), which is discussed in the next chapter.

12 In contrast to this, and as will be revealed in the following chapters, Hughes's children's poetry and prose go beyond the stage of enumerating dualities in order to successfully unify them as part of his healing quest.

13 The influence of Beckett can also be found in *Wodwo* (1967) and *Crow* (1970), both of which can be considered as grotesquely Beckettian. *Wodwo* recalls a dark world, corresponding to the wasteland of modern Christian civilisation, very much like the world of Beckett. *Crow*, written in a mixture of deliberate crudity and absurdity and set halfway between apocalypse and surrealism, draws strongly on myth and folklore, as do Hughes's children's plays. However, in the case of *Crow*, Hughes's resource to myth and folklore, allow him to mock God's creation and deconstruct the Christian creation myth in order to reconstruct his own personal myth. Crow is employed by Hughes as a trickster figure, and during his wanderings is accompanied by a fool, whom he meets in the desert waiting by a single tree, obviously recalling *Waiting for Godot*. Moreover, Crow's dismemberment suggests that he is also conceived as a shaman, and is a vital part of Hughes's adult healing quest, although unlike his children's works, in *Crow* redemption cannot be fully obtained.

14 Depth Psychology is one of the many therapeutic approaches derived from Jungian Psychology and refers to what lies beneath the surface, at the depth of our experience. The method relies on discovering the motives behind mental problems as a means to treat them and seeks to restore the individual's connection to the Self.

15 In the tale, a spoiled princess befriends the Frog Prince who is transformed into a handsome prince. However, unlike 'Beauty and the Beast' it is not compassion that reverses the curse and transforms the monster into a human, but an act of passion, as she hurls the frog against the wall.

16 There is, of course, a rich tradition of motifs regarding the choice of roads and crossroads in both folklore and literature. Amongst the ancient Greeks, crossroads were associated with gods representing the underworld or darkness. Similarly, handbooks of magic from the Medieval Age mention crossroads as a place to conjure the devil, and in folklore and myth magical rites supernatural creatures are often associated with crossroads.

17 A longer analysis of 'Orpheus' appears in Kerslake, 'Constructing and Deconstructing the Complexities of Orpheus in Ted Hughes's Healing Quest', in *The Politics of Traumatic Literature: Narrating Human Psyche and Memory* (Newcastle upon Tyne: Cambridge Scholars, forthcoming 2018).

18 Hughes had already dedicated *Crow* to the memory of both Assia Wevill and Shura.

19 Incidentally, after Plath's death Hughes became her literary executor, overseeing future editions of her work.

20 According to Virgil's version, Orpheus was later killed by the women of Thrace. The reason and manner of his death vary in different accounts, but according to the earliest account, that of Aeschylus, they were Maenads urged by Dionysus to tear him to pieces in a Bacchic orgy because he preferred the worship of the rival god Apollo. However, miraculously, as Orpheus is dismembered and his head and lyre float down the river, he continues to sing.

References

Aizenstat, Stephen, 'Jungian Psychology and the World Unconscious', in Theodore Roszak, Mary E. Gomes and Allen D. Kanner (eds), *Ecopsychology: Restoring the Earth Healing the Mind* (San Francisco: Sierra Club Books, 1995), pp. 92–100.

Aizenstat, Stephen, 'Nature Dreaming: Depth Psychology and Ecology', 2003. Web. 15 March 2012. www.dreamtending.com/naturedreaming.pdf.

Beckett, Samuel, *The Complete Dramatic Works* (London: Faber & Faber, 2006).

Bettelheim, Bruno, *The Uses of Enchantment* (New York: Vintage Books, 1977).

Carson, Rachel, *The Sense of Wonder* (New York: Harper & Row, 1965; New York: HarperCollins, 1998).

Chaudhuri, Una, '"There Must Be a Lot of Fish in that Lake": Toward an Ecological Theater', *Theater*, 25:1 (Spring/Summer 1994), pp. 23–31.

Cobb, Edith, *The Ecology of Imagination in Childhood* (New York: Columbia University Press, 1977).

Doolan, Moira, 'Living Language', BBC pamphlet teachers' notes, edited by Joan Griffiths (London: British Broadcasting Corporation, Autumn 1974).

Graves, Robert, *The White Goddess: A Historical Grammar of Poetic Myth* (London and Boston: Faber & Faber, 1948; New York: Farrar, Straus & Giroux, 1966).

Griffiths, Joan and David Kerrison (eds), 'Living Language', BBC pamphlet teachers' notes (London: British Broadcasting Corporation, Autumn 1977).

Haase, Donald, *The Greenwood Encyclopedia of Folktales and Fairy Tales* (Connecticut, CT: Greenwood Press, 2008).

Hong, Chen, 'Hughes and Animals', in Terry Gifford (ed.), *The Cambridge Companion to Ted Hughes* (Cambridge: Cambridge University Press, 2011), pp. 40–52.

Ted Hughes, 'The Importance of Being Earnest', in J. E. Fisher (ed.), *The Don & Dearne, being the Magazine of Mexborough Grammar School*, July 1948, p. 15.

Hughes, Ted, *Nessie the Mannerless Monster* (London: Faber & Faber, 1964).

Hughes, Ted, *Poetry in the Making* (London: Faber & Faber, 1967).

Hughes, Ted, *Wodwo* (London: Faber & Faber, 1967).

Hughes, Ted, 'Myth and Education', *Children's Literature in Education*, I (March 1970), pp. 55–70.

Hughes, Ted, *The Coming of the Kings and Other Plays* (London: Faber & Faber, 1970).

Hughes, Ted, *The Tiger's Bones* (New York: The Viking Press, 1974).

Hughes, Ted, *Shakespeare and the Goddess of Complete Being* (London: Faber & Faber, 1992).

Hughes, Ted, 'Foreword', in *The Sacred Earth Dramas* (London: Faber & Faber, 1993), pp. vii–viii.

Hughes, Ted, *Winter Pollen* (London: Faber & Faber, 1994).

Hughes, Ted, *Birthday Letters* (London: Faber & Faber, 1998).

Hughes, Ted, *Alcestis* (London: Faber & Faber, 1999).

Hughes, Ted, *Collected Plays for Children* (London: Faber & Faber, 2001).

Kahn, Peter, *The Human Relationship with Nature: Development and Culture* (Cambridge, MA: MIT Press, 1999).

Kerslake, Lorraine, 'Constructing and Deconstructing the Complexities of Orpheus in Ted Hughes's Healing Quest', in *The Politics of Traumatic Literature: Narrating Human Psyche and Memory* (Newcastle upon Tyne: Cambridge Scholars, forthcoming 2018).

Manes, Christopher, 'Nature and Silence', in Cheryll Glotfelty and Harold Fromm (eds), *The Ecocriticism Reader: Landmarks in Literary Ecology* (London: University of Georgia Press, 1996), pp. 15–29.

Marchenkov, Vladimir, *The Orpheus Myth and the Powers of Music* (Hillsdale, NY: Pendragon Press, 2009).

May, Theresa, 'Greening the Theatre: Taking Ecocriticism from Page to Stage', *Interdisciplinary Literary Studies: A Journal of Criticism and Theory*, 7:1 (2005), pp. 84–103.

May, Theresa, 'Kneading Marie Clements' *Burning Vision*', *Canadian Theatre Review*, 144 (Fall 2010), pp. 5–12.

Metzner, Ralph, 'The Psychology of the Human–Nature Relationship', in Theodore Roszak, Mary E. Gomes and Allen D. Kanner (eds), *Ecopsychology: Restoring the Earth Healing the Mind* (San Francisco: Sierra Club Books, 1995), pp. 55–67.

Nicholson, Max, *The Environmental Revolution* (London: Hodder & Stoughton, 1970).

Plath, Sylvia, *Letters Home: Correspondence 1950–1963*, edited by Aurelia Plath (London: Faber & Faber, 1976).

Powell, Barry B., 'Myths of Death: Encounters with the Underworld', in *Classical Myth* (Upper Saddle River, NJ: Pearson Prentice Hall, 2004), pp. 288–317.

Reddick, Yvonne, *Ted Hughes: Environmentalist and Ecopoet* (Basingstoke: Palgrave Macmillan, 2017).

Reid, Christopher (ed.), *Letters of Ted Hughes* (London: Faber & Faber, 2007).

Roszak, Theodore, Mary E. Gomes and Allen D. Kanner (eds), *Ecopsychology: Restoring the Earth Healing the Mind* (San Francisco: Sierra Club Books, 1995).

Sagar, Keith (ed.), *The Art of Ted Hughes* (Cambridge: Cambridge University Press, 1975).

Sagar, Keith (ed.), *The Laughter of Foxes* (Liverpool: Liverpool University Press, 2006).

Sagar, Keith (ed.), *Ted Hughes and Nature: 'Terror and Exultation'* (Peterborough: Fast Publishing, 2009).

Sagar, Keith (ed.), *Poet and Critic: The Letters of Ted Hughes and Keith Sagar* (London: British Library, 2012).

Schofield, Annie, 'Hughes and the Movement', in Keith Sagar (ed.), *The Achievement of Ted Hughes* (Manchester: Manchester University Press, 1983), pp. 22–36.

Ward, Donald, 'Idionarrating and Social Change', in Röhrich Lutz and Sabine Wienker-Piepho (eds), *Storytelling in Contemporary Societies* (Tübingen: G. Narr Verlag, 1990), pp. 33–42.

Warner, Marina, *From the Beast to the Blonde: On Fairy Tales and Their Tellers* (London: Chatto & Windus, 1994).

White, Lynn, Jr., 'The Historical Roots of Our Ecologic Crisis', *Science*, 155 (10 March 1967), pp. 1203–1207.

Wilson, E. O. and Stephen R. Kellert (eds), *The Biophilia Hypothesis* (Washington, DC: Island Press, Shearwater Books, 1993).

Newspaper

J. E. Fisher (ed.), *The Don & Dearne, being the Magazine of Mexborough Grammar School*, July 1947, pp. 13–14.

5 Hughes's children's poetry

Breaking down barriers: children's poetry according to Hughes

Hughes's poetry falls into three categories: that, strictly speaking, for children (such as *Meet My Folks* (1961) or *The Cat and the Cuckoo* (1987)), his adult poetry (from *Crow* (1970) to *Birthday Letters* (1998)) and a more extensive category accessible to both adults and children (i.e. *Season Songs* (1976)). This brings us to the inevitable question and ongoing critical discussion of what constitutes children's poetry as opposed to adult's poetry.

Already in 1963, W. H. Auden set the precedent by rightly claiming that 'while there are some good poems which are only for adults, because they presuppose adult experience in their readers, there are no good poems which are only for children' (181). Morag Styles takes Auden's argument one step further by contending that: 'there is no such thing as poetry for children' (2004: 396).

So, in drawing a line, if one can be identified, between adult and so-called children's poetry, the differences would appear to lie mainly in elements such as subject matter, word choice and complexity. Andrew Motion, who was appointed Poet Laureate in 1999, following the death of Hughes, has expressed perhaps better than anyone, why poetry matters in children's lives and what the pleasure of poetry consists in: 'it allows children to feel a profound sense of connection with their interior spaces, and also to make links with the wider world which lies around them' (2010: ix).

Hughes's own opinion of what constitutes children's poetry would appear to be similar to that of both Auden and Styles, and very much in line with Motion. Hughes himself eloquently conveys how the problem of writing poetry for children is that:

> we think children need a special kind of poetry. Each writer for children has his own idea of what that is … Publishers, of course, know that poetry is not sold to children – it is sold to their parents or teachers. So this is the barrier to publishing children's poetry. The author thinks he knows what they want, or need, the teachers or parents think they know best. And the publisher thinks he knows best what poet & teacher think they know best.

> And we all think differently. Each author writing for children writes the same – and all write differently.
>
> (In Paul, 2005: 257)[1]

According to the poet, within these publishing constraints, the problem of writing poetry for children lies in the problematic that surrounds its audience: who is the text for? Who buys it? Who reads it? Hughes thought that there were no fixed boundaries between writing for children and writing for adults, and believed that such barriers were artificial. He attempted to overcome this drawback, by communicating in an open and direct way in order to reach both the adult/child reader. As a result, many of his poems for children have appeared in collections for adults and vice versa.[2] The suitability of Hughes's poetry for children is a question that will be reviewed later, given that certain books would appear to cross the boundaries of the child/adult reader more successfully than others.

As previously stated, Hughes's ideas of childhood and of how poetry should be written are very much in line with the Romantic vision of Blake and Wordsworth, especially concerning his views on imagination, which are explored in *Poetry in the Making* (1967). In the Introduction, Hughes claims that an integral part of his method is not '"How to write" but "How to try to say what you really mean" – which is part of the search for self-knowledge and perhaps, in one form or another, grace' (*PM*: 12). Hughes's search for self-knowledge is part of his own quest for the truth. Hughes had a clear idea of what he wanted to achieve throughout his children's writing, and especially in the case of his poetry: to write in a lingua franca, or as he puts it, in 'a style of communication for which children are the specific audience, but which adults can overhear' (*LTH*: 482) as long, that is, as they 'suspend their defences and listen – in a way secretly – as children. So long as the affection is there' (*LTH*: 482). Hughes's careful choice of the words 'affection' and 'grace' highlight the enormous potential that Hughes felt lay within children. By 'affection' Hughes appears to be referring to a desire to communicate directly to children, in a way that is both challenging and unpatronising. Another element that comes out of his lingua franca is the fact that not only was he writing for children within hearing of adults, but as the poet himself remarked in the composition of *Season Songs* (1976) he was trying to write in a language that was not childish, but one that was 'within the hearing of children'. What remains clear is that much of Hughes's children's poetry is not poetry about childhood or even poems exclusively for children as such, but are poems written with 'affection' which can reach both audiences.

In Hughes's poetic universe, animals, the countryside, nature and conservation are all constant themes. His poetry contains a profound sense of respect and responsibility towards a threatened and diminishing natural world, one where, of course, animals abound. It is through animal imagery that Hughes exalts the instinctive power of nature that he finds lacking in human society. Moreover, the poet's interest in the natural world and his engagement with nature helped him to capture real animals through words. As he puts it, the secret of good poetry is knowing how to make the poem come *alive*: 'Imagine what you are writing about. See it and live it.... When you do this, the words look after themselves, like magic' (*PM*: 18). In that way, the poet tells us, 'You will have captured a spirit, a creature' (*PM*: 18).

Much of Hughes's poetry is built on this premise. His poems about animals have the ability not only to awaken our senses and capture the essence that lies within the poem, but the poem itself appears to be *endowed* with life. Hughes poetic voice reminds us that we are 'both a part of and apart from nature' (Bate, 2000: 33). Given that the alienation of human society/nature is central to Hughes's entire poetic quest, healing the rift between the human and 'more than human' world[3] is a central feature of his children's poetry, which can be read as the chance for adults to recover their senses and reconnect with nature.

Perhaps no one has written more successfully on the alienation of humans and animals than John Berger. In his influential essay 'Why Look At Animals?' Berger argues that industrial capitalism has transformed human's relationship with the natural world, so that when animals and humans look at each other it is always across a gap of non-comprehension. The human gaze has objectified animals so that 'animals are always the observed.... What we know about them is an index of our power, and thus an index of what separates us from them. The more we know, the further away they are' (1980: 16). Randy Malamud takes Berger's question to the next level by looking at how human culture views animals and comments on the barrier that separates the human from the non-human, 'The relationship between people and nonhuman animals is codified in social culture as hierarchical and fundamentally impermeable: we are in here, they are out there' (2003: 3). Gary Kowalski also refers to this awareness of animals that Hughes so gracefully conceives. As Kowalski asks: 'What do we see when we look into the eyes of another living creature?' (1991: 91). By questioning whether we see animals as they really are, Kowalski enhances our awareness and cautions us to pay attention, reflecting on what we see, and more importantly, 'to what you cannot see ... What you look upon is a living spirit. Greet and respect it. Appreciate it for what it is. Ask yourself: What does it feel like to be this creature?' (1991: 91).

Like both Berger and Kowalski, when Hughes writes about animals he has the ability (and poetic sensitivity) to capture the animal's gaze and bring the poem to life. This is perhaps best illustrated in Hughes's much celebrated and first animal poem 'The Thought Fox',[4] a metafictional poem about writing a poem about an imaginary fox. The fox becomes alive as the poem progresses: it becomes real both in the poet's mind, within the poem and at the same time through the reading process, in the imagination of the reader. This is achieved through the poet's description of the fox and its movements, to the point that there is a 'sudden sharp stink of fox' and 'The page is printed' (*NSP*: 3). Not only can the reader *see* the fox, but *feel* him and even *smell* him. He is now a true, tangible creature that exists not only in the mind of the poet but in the mind of those who read the poem. For, as the poet so elegantly puts it, his poems are 'a sort of animal. They have their own life, like animals' (*PM*: 15).

For Hughes, the animal enters the poem through the poet's observations. His poetic voice is primarily shamanic and poetry, for Hughes, was the means to transport the human spirit from the dark, subconscious side of human nature into the world of light. The creatures in his poems are totem beasts. His poems can be seen as regenerations of spirits, cures and celebrations of life and death. The poet's shamanic quest is to discover and recover the 'truth', that we are part of nature, through healing images, in order to regenerate and reconnect. Whilst in his adult poetry these tensions are not fully resolved, hence the poet appears wounded and broken – through images of death, destruction and annihilation – as part of the symbolic dismemberment of his shamanistic journey, it is in his children's poetry that the shaman/poet tries to find a balance between the chaotic energies of the universe, especially those of humankind and nature. As part of his healing quest, his children's poetry allows him to bring these dualities into dialogue. In contrast with his children's plays in which he sets up the oppositions between our inner/outer worlds and creative/destructive universes, his children's poetry goes a step further, seeking to reconnect dualities, connecting the outer world (that of our surroundings) with the inner world (that of our senses).

Hughes's ability to 'animate' his poetry by capturing so vividly the animal's gaze and depicting it with such grace connects to what Jonathan Bate calls an 'ecopoetic consciousness' (2000: 42). Bate argues for poetry as a vital and necessary element of reimagining human relationships with the natural world and explores the idea of ecopoetics as 'a description of dwelling with the earth, not a disengaged thinking about it, but an experiencing of it' (2000: 42). Bate, together with Jonathan Skinner, both emphasise the domestic space in the root of 'eco'[5] in the term ecopoetics as a making, through poetry, of the dwelling place or home, so that

> 'Eco' here signals – no more, no less – the home we share with several million other species, our planet Earth. 'Poetics' is used as *poesis* or making, not necessarily to emphasize the critical over the creative act (nor vice versa). Thus ecopoetics: a house making.
>
> (Skinner, 2001: 7)

Drawing on J. Scott Bryson's definition that 'ecopoets offer a vision of the world that values the interaction between two interdependent and seemingly paradoxical desires, both of which are attempts to respond to the modern divorce between humanity and the rest of nature' (2002: 8), the term 'eco-poet' appears more suitable than 'nature poet' to refer to Hughes since his poetry goes beyond traditional Wordsworthian nature poetry by investigating the relationship between nature and culture. Not only does he address problems related to the environmental crisis but his poetry reflects this concern for the more-than-human world by 'putting our inner worlds in contact with the outer world' (Gifford, 2011: 9).

Ed Roberson's definition of the term is particularly useful to illustrate Hughes's sense of poetic responsibility. As he explains, ecopoetry is not simply about 'the experience of writing as it is about the experience and observation of nature' (Roberson, 2009: 3). It is precisely this definition of ecopoetry that Hughes enacts in his children's poetry and which is key to his healing quest, by seeking to reconnect the dualisms that have alienated humans from nature and the sense of being part of the web of life.

Given the daunting weight and scope of Hughes's poetic oeuvre, this chapter surveys the development of Hughes as an eco-poet and analyses the development of his children's poetry. It follows a chronological analysis, divided into three parts, to look at the different stages in Hughes's poetry focusing on those poems which best illustrate Hughes's healing quest. First, it looks at the poetry of his early years and the first two collections of Hughes's children's poetry, *Meet My Folks* (1961) and *The Earth Owl and Other Moon People* (1963), together with *Nessie the Mannerless Monster* (1964) as an example of poetic prose. It then analyses Hughes's most important poetry collections for children (from the mid-1970s to mid-1980s) including *Season Songs* (1976) and *Under the North Star* (1981), and draws on the importance of visual art to Hughes, and his artistic collaboration with Leonard Baskin. Finally, it looks at Hughes's less accredited collaboration with Reg Lloyd in the 1980s, beginning with *What is the Truth?* (1984) and closing with two of Hughes's last books of poetry written for younger children: *The Cat and the Cuckoo* (1987) and *The Mermaid's Purse* (1993), both of which have often been overlooked and rarely acknowledged by existing criticism.

The seeds of Hughes's inspiration: from *Nessie* and *Meet My Folks*, to Hughes's moon poems

Nessie the Mannerless Monster *(1964)*

The previous chapter highlighted that one of the main characteristics of Hughes's children's plays was the fact that many of them were written as dramatic verse or characterised as poetic prose. As a bridge between Hughes's children's plays, his poetry and his prose *Nessie the Mannerless Monster* serves as a singular example of Hughes's poetic prose.

The actual story behind *Nessie* also deserves a brief mentioning. In 1959 Tim Dinsdale, one of the most dedicated searchers and prolific authors of the Nessie phenomenon, began making a lengthy, detailed analysis of eyewitness reports. In April 1960, Dinsdale achieved a moving film of what believers agree is Nessie's hump. He informed the British Museum, who suggested he should avoid spreading the news, however by June his film had been shown on television, arousing great public interest. He later began writing his book, *Loch Ness Monster*, which was published in 1961. Hughes wrote a review of Dinsdale's book which was published in *The New Statesman*, that same year. Hughes's analysis is quite remarkable:

> Call a thing 'monster' and it disappears, submerging into a dream world among angels and black gods ... So even if the Loch Ness Monster is 60 feet long, weighs several tons, and is one of a fairly well-distributed species, breeding true, it will never get its official investigation till we call it something else.
>
> (1961: 61)

The fact that Dinsdale's book inspired Hughes appears, to this date, to have been overlooked by critics. On examining the dates there is however little doubt that the book was the genesis behind Hughes's poem 'Nessie'. Plath wrote in a letter to her mother that 'Ted has written a lovely poem about the Loch Ness Monster for *Vogue* (a children's poem)' (1976: 433). Indeed, curiously enough the poem first appeared in *Vogue*, in December 1961, and was later to become Hughes's third picture book for children, *Nessie the Mannerless Monster*, first published 24 April, 1964, with pictures by Gerald Rose. The British Library holds a manuscript with an early draft of the poem 'Nessie', which varies slightly from the later published version. In it Hughes talks of Nessie as a 'rejected monster, a monster without protection'.[6] However, in the published edition the monster takes on a female identity, 'She is sick with sorrow', for 'Nobody thinks I exist' (1964: unnumbered). In *Poetry in the Making* (1967), Hughes gives us a possible clue as to why the monster should be female, by reminding us that, 'In olden times, the monster used to be worshipped as a goddess, and was called Nessa, so it is quite curious that she should still be called Nessie' (*PM*: 108). His commentary about returning to the past, to what this ancient world was once like, reveals a much deeper concern, and one that runs through a great deal of his poetry: what we are today, how we live, owes something to this lost world, one in which he believes the true essence of humanity lies.

That Hughes chooses here to focus his tale on a 'monster' rather than a charismatic mammal, a pet or even a bird is also telling. From an ecocritical perspective the figure of the animal has played an important role throughout Western history and literature (Aftandilian, 2007; Ham and Senior, 1994; Kalof, 2007; Midgley, 1984). The making of the boundaries which separate the human from the non-human, and the way in which human culture has viewed animals, have long been regarded as issues of central importance to ecocritics. In view of this, *Nessie the Mannerless Monster* can also be read as an ecological moral tale. Hughes chooses to blur and overlap the human/animal boundaries focusing not so much on animal differences here but rather on their similarity to humans, especially by using the figure of the speaking animal to appeal to his children's audience.

This is portrayed in the tale when Nessie goes to London and a group of 'Ban the Bomb' protestors want to adopt Nessie, a 'peaceful beast, beautiful and dumb' (1964: unnumbered), as their mascot. However, on arriving at Trafalgar Square all the people disappear, leaving Nessie abandoned. After being rejected by London zoo, she is sent to Kensington Museum,[7] where all her cousins are 'prettily encased'. However, once there Hughes's Nessie receives the same dismissal as Dinsdale's monster. Despite insisting that she is 'not a fairy-beast and not a dream-beast', the sceptical scientists at the museum, blinded by their own knowledge, insist that she is an imposter since all the plesiosaurs have been dead for millions of years. The only person able to recognise Nessie, is the Scots poet Willis, who despite being penniless 'knew a monster when he saw one' (1964: unnumbered). Willis is the only person whose imagination remains intact, and proves in his quest that poetry can triumph over science. He takes Nessie to the palace to meet the Queen, who touched by his story appoints him 'Vice-Regent of Scotland, Wales and Northern Ireland', suggesting perhaps that England is beyond redemption, and Willis becomes a civil servant and looks after Nessie making sure 'that she's well fed'.

Most critics have perhaps unfairly accused the narrative of being slack, written in uneven doggerel, imitating the style of the Scots poet William McGonagall (Cushman, 1983: 242) and in the somewhat erroneous words of Michael Lockwood, would 'certainly sound inept to the adult reader' (2009: 299). Perhaps Lockwood is missing the obvious, that Hughes was not writing for adults here? Although the picture book is not Hughes's best, critics appear to have overlooked the interest of the story and the fact that it can be read as the genesis of the development of his ideas on the role of imagination and myth, used as creative tools to restore the balance between the inner and outer worlds of Western society, depicted here as the underwater world and the other. The conflict between creativity and rational science which Hughes introduces here for the first time is also strongly reminiscent to that of Dully and the master in his children's play 'The Tiger's Bones' (1965). Much of the argument behind the narrative draws on Hughes's critique of contemporary culture and society, ideas which he would later develop in his essay 'Myth and Education' (1976 in *WP*).

Meet My Folks *(1961)*

Meet My Folks was Hughes's first children's book, dedicated to his daughter Frieda. In a letter to Leonard Baskin, written in July, 1959, Hughes mentions that he had written a book of poems about relatives, commenting that 'American publishers are shy of it – all the characters being metamorphotic or psychotic' (*LTH*: 147). Despite initial reticence to put it into print, the book was published by Faber in 1961, along with George Adamson's intensely detailed illustrations.[8]

Curiously *Meet My Folks* (1961) is the only collection that Hughes wrote about people. Hughes was, after all, a poet of nature and not of people, capturing animals and bringing them to life with more skill and grace than he did his imaginative relatives. In the chapter 'Meet my Folks', published in *Poetry in the Making* (1967), Hughes plays with the idea that the problem with relatives is that 'you are stuck with them, whether you like them or not. And they seem to think they own you as if you were their pet cat' (*PM*: 103). He goes on to suggest a way to reinvent them by rearranging them under different names and appearances. Amongst these characters is an uncle who invents things, a grandfather who traps owls, a grandmother who is an octopus, his brother Bert who 'used to go to school with a Mouse in his shirt' (*CPC*: 45) a father who works as the Chief Inspector of holes and a fairy godmother who saves him from 'The Wicked Powers' (*CPC*: 40).

In *Written for Children* (1983), author and academic John Rowe Townsend claims that the book

> reflects the tendency of its decade to break old moulds; it gets well away from the charming harmlessness long associated with children's verse, and describes a family you might not particularly want to meet, still less to acknowledge for your own.
>
> (305)

Neil Roberts claims that it is written in 'rollicking, heavily accented verse interestingly reminiscent of Hughes's juvenilia derived from Kipling and Robert Service' (2006: 168). However, Roberts fails to mention the more obvious similarities with Edward Lear's nonsense verse, such as the evasive and often absurd tone or the undercurrent of violence used in a comic way to delight children, as in 'the cat split laughing and had to be stitched' (*CPC*: 38). Cushman informs us that 'Hughes stretches the boundaries of the child's imagination by transforming a family into a wild, strange collection of eccentric persons and odd creatures' only to conclude that 'the problem with *Meet My Folks* is in the poetry' (1983: 241). As if it to justify himself Cushman then goes on to quote Sagar's observation that 'Few lines have the inevitability that would make them memorable. The poems are fresh and engaging, but shapeless' (1983: 241).

Perhaps Sagar was referring here to the fact that Hughes's first book of poetry was heavily influenced by the renowned American children's writer and illustrator Dr. Seuss, whose picture books remained extremely popular in USA. Hughes himself would have been familiar with *The Cat in the Hat* (1957) as well as *Green Eggs and Ham* (1960). Moreover, whilst living with Plath, in USA, Hughes reviewed several of Dr. Seuss's books, including *The Cat in the Hat Comes Back* (1958) and *One Fish Two Fish, Red Fish, Blue Fish* (1960). The following extract from 'My Brother Bert' could have easily been written by Seuss, and illustrates how Hughes draws on his silly sense of humour and word play, as well as his characteristic and rather clumsy rhyming style:

> What if the Pangolins were to caper
> Out of their nests behind the wallpaper?...
> And out of a shoebox the excitable Platypus
> Along with the Ocelot or Jungle-Cattypus?
>
> (*CPC*: 45)

Indeed, it would appear that most of Hughes's critics have been so busy looking for aspects to criticise that they appear to have overlooked a crucial aspect of the poet's creative work here: the fact that this is the first time that Hughes introduces the crow as a trickster figure.[9] 'My Sister Jane' is not just 'a bird' but 'a great big crow' (*CPC*: 39), and despite her human appearance with her dark glasses and curly wig she behaves like a real crow.

Hughes also uses elements of folklore and fairy tale, reminiscent of *Sleeping Beauty* and *Jack and the Beanstalk* such as in his 'My Fairy Godmother' poem. The poem is about a magic ladder that allows the protagonist to travel between two worlds. Already, right at the beginning of Hughes's shamanic quest, the ladder can be seen as a symbol to negotiate between the forces of the universe. According to Eliade (1964: 265), the shamanic ladder is the earliest version of an idea of an axis of the world, also known as the World Tree, which is correspondingly the Tree of Life in Kabbalism. The axis mundi is the connecting point between the different levels of the cosmos, of heaven and earth, and can be found in numerous creation myths (Pratt, 2007: 436). This idea of the axis mundi as a connecting point between different worlds is one that can be found elsewhere in Hughes's poetry. One of the best poems to illustrate this is perhaps 'My Own True Family' which closes the collection and was significantly added 12 years later. Whilst Roberts calls it 'a shamanic narrative' (2006: 169), even Cushman considers that it 'aspires to the condition of myth' (1983: 242).

Hughes's interest in folklore and myth can be seen in the poem which plants the seed for Hughes's healing quest and draws on the idea of the 'Tree of Life'. It opens with a child in an oak wood looking for a stag. He is confronted by an old woman who puts him under a spell in which the trees identify themselves as 'his own true family'. They make him promise that 'Whenever you see an oak-tree felled, swear now you will plant two' (*CPC*: 53). Compared to his other poems, 'My Own True Family' has a much more serious tone and reads as a prophetic journey with the boy returning from his dream, altered by the experience. It can be read as one of Hughes's first children's poems of regeneration and healing, as a kind of ecological parable, in which the oak symbolises Hughes's healing quest to reconnect human beings with nature. However, the poem could also be considered as the genesis behind Hughes's later tree-growing project with schools in the 1970s.

In a letter Hughes wrote on 9 November 1972, to Caroline Moorehead, Features Editor for the *Times Educational Supplement* he gives details of his plan, claiming that the idea was to involve children by enacting a sense of responsibility in them by showing them that the trees belonged and depended on them.[10] The letter appears to echo the poetic statement of the poem, which introduces Hughes's notion of interconnectedness of the human and non-human. The last lines of the poem echo the transformation the child has undergone after being in the oak wood and reconnecting with nature, to the point that 'My walk was the walk of a human child, but my heart was a tree' (*CPC*: 53). Hughes was already laying the foundations here for a more holistic view of nature anticipating Carson's sense of wonder by realising that the path to build a human consciousness of nature lies in responding to the wonder expressed by children.

The Earth Owl and Other Moon People (1963)

The Earth Owl and Other Moon People was published the same month as *How the Whale Became*, in November 1963. Sylvia Plath had taken her own life less than nine months before leaving Hughes to raise their two children, with a profound sense of guilt, as can be seen in various letters that he wrote in the days following the tragic event, 'I was the only one could have helped her, and the only one that couldn't see that she really needed it this time. No doubt where the blame lies' (*LTH*: 214).

As previously mentioned, Hughes's estrangement from his adult work, from the time of Plath's suicide in 1963 until the publication of *Wodwo* (1967), followed by *Crow* (1970), and his creative surge as a children's author following Plath's death was not simply a coincidence, nor was it only an outcome of having to raise two young children without a mother. There is, of course, an element of necessity behind his children's work, and one which is seldom mentioned: the economic factor. Not only did Hughes write more books for children than collections of original poetry for adults, but the children's books had longer print runs. When Hughes wrote to his brother Gerald as early as August 1957, claiming that 'if I can keep this up … then that will be quite a wage' (*LTH*: 108), he was well aware of the potential income that lay in children's writing. Notwithstanding, his children's writing was an important therapeutic way for Hughes to deal with his terrible loss.[11] The adversity of these circumstances can be seen in 'Moon-Ways' where the poet tells of 'The road that takes you home loyally/Year after year, suddenly cruelly/Runs you to jail instead' (*CPC*: 83).

Meanwhile, another project that was in full progress at the time, and which received wide coverage in the media, was the space race between Russia and USA. Incidental or not, Hughes produced more than 50 moon poems, published in a succession of volumes between 1963 and 1988.[12] In *Poetry in the Making* (1967), Hughes dedicated an entire chapter to the creative universe of his moon world and the strange creatures that dwell there, claiming that 'the number and oddity of the creatures which inhabit the earth or the planets, are nothing to those which inhabit our minds' (*PM*: 110). Indeed, the poet's moon poems present us with an eerie haunting moon world, offering us a glimpse of his own creative universe and presenting us with surreal, extreme landscapes of imagined nightmarish creatures that inhabit parallel universes. The poem 'Moon Walkers', from the *Moon-Whales* (1976) collection, is about the nightmare creatures that live inside our dreams. A bad night's sleep together with the full moon set the scene for the monsters that dwell inside our darkest nightmares. The setting is strongly evocative of Maurice Sendak's renowned picture book *Where the Wild Things Are* (1963), written for a younger audience, and published the same year. Both authors confront our inner world, and the things that exist in our imagination, exploring the fears, desires and nightmares that lie below the surface.

In the debate about what constitutes poetry for children, certain children's literary critics have suggested that Hughes's poetry is too difficult for children.

When Hughes was awarded the first *Signal* Poetry Award for *Moon-Bells* (1978), it was considered a controversial choice. According to John Wain, 'None of this is poetry for children' (1979: 65). In the next issue of *Signal* Stephen Corrin took up his comment by saying that Hughes's poetry is 'outside the strictly limited spiritual experience of the child' (1979: 169). All these comments suggest a patronising attitude to children's poetry and support the idea that a poet who writes for children must write down to them, ideas that Hughes, of course, resisted and argued against.

Some critics would appear to find Hughes's moon poems unsuitable for young readers. John Adams has criticised his dark poetic vision claiming that 'whether for children or adults, this is obsessionally narrow and heavy-handed, and in a way sadistic' (1980: 107). He goes on to state that whilst Hughes promotes creativity in children's writing 'he shows little sympathy for humanity and tends towards a narrower and blacker view of life to the point of chaos' (Adams, 1980: 107). Meanwhile Anthony Wilson claims that *Moon-Whales* (1976) is 'one of Hughes's most consistently dull and predictable books' and that reading it 'is a bit like being hit over the head repeatedly' (2001: 81).

Likewise, the majority of Hughes's critics have not given positive feedback to his moon world either. Sagar's comments are not exactly flattering; 'Though he tries very hard to be grotesque and clever, Hughes cannot breathe life into his moon creatures' (1972: 15–16). Certainly poems such as 'Moon-Witches', 'The Moon-Hyena' or 'Moon-Hops' are not the poet's best. Their loose rhyme is often unconvincing and gives the impression of being clumsy. However, despite the gory details, Hughes's treatment of violence involves fantastic creatures which are so bizarre that it remains removed from the everyday life of the child reader. In 'Moon-Horrors' Hughes describes the gory fate of the victims of the different moon-creatures including a man-eating tiger, a shark or mosquitoes who 'drink at a man as he sleeps' (*CPC*: 78). Some of the creatures are so horrible that they have no name, just a number, such as the 'gross number three' (*CPC*: 78) who 'attacks as a nightmare' (*CPC*: 78).

There are of course exceptions: 'Moon-Theatre' and 'Moon-Marriage' offer an interesting insight into Hughes's imaginative world and are rich with layers of meaning. Both poems show the lunatic influence of the moon over human imagination. 'Moon-Theatre' describes a daydream which sets the scene for the shaman's possession of a spirit guide: 'Then tap a drum and fix your eyes in a glassy stare' (*CPC*: 73) and tells the story of a captive princess, abducted by an ogre, who 'must howl and rave' until she escapes disguised in a wolf's skin. 'Moon-Marriage' describes the process of being married in which a totemic animal in the form of a frog, or 'maybe a smiling wolf' (*CPC*: 94) chooses a bride. As Mick Gowar has pointed out, the collection is 'concerned with the notion of crossing thresholds, of inhabiting, and occasionally being trapped, within boundaries and borderlands' (2011: n.p.). Indeed the characters, monsters and fairy tale creatures are those that may lurk at the bottom of our own dreams, or nightmares, crossing the boundaries between life and death. It is through these poems that Hughes crosses over and breaks down barriers, exploring the fears of our own inner psyche, so that the 'irrational' crosses the borders of fact/fiction, child/adult, violence/beauty showing us the truth of the dark side of the moon.

Most of Hughes's critics would appear to have overlooked one of the fundamental aspects of Hughes's moon poems. Claas Kazzer has suggested that *Earth Moon* 'presents the laws of the Goddess, who appears more threatening the more alienated she has become. And it represents the dark and light of the "female side" of his own inner being' (2004: 119). Given that Hughes specifically connects his inner moon world with poetry, it would appear that is through his moon poems that the poet offers the world of the moon goddess to us, and as Skea has suggested (2012), he uses alchemy in his poetry as a way of healing the divisions and crossing borders. Furthermore, given the context in which they were written, not only did Hughes dedicate his first collection of moon poems to his children, but he would have read them the poems at bedtime. Given the underlying sad, but deeply affectionate, tone of many of the poems, instead of suppressing emotions, they would appear to have had a special consoling power both for Hughes and his children serving as a way to deal with their own pain and mourning.

The moon goddess can also be read here from an ecofeminist perspective, which can also be applied to Hughes. According to Carolyn Merchant (1992) cultural ecofeminism often stems from an anti-scientific or technological viewpoint (which would certainly apply to Hughes). As she writes, it 'celebrates the relationship between women and nature through the revival of ancient rituals centered on goddess worship, the moon, animals, and the female reproductive system' (Merchant, 1992: 191). Like in Hughes's vision, 'nature is held in esteem as mother and goddess is a source of inspiration and empowerment' (Merchant, 1992: 191).

As previously mentioned, Hughes's goddess of nature was clearly inspired by Graves's *The White Goddess* (1948), and draws on Graves's depiction of the nature goddess in her three aspects of maiden, mother and crone. In her book *The Spiral Dance* (1979) Starhawk, perhaps best known as pioneer in the revival of earth-based spirituality and goddess religion, also comments on the triple goddess portrayed in her different expressions:

> The Maiden aspect of the Goddess shows women how to be independent and strong; the Mother aspect shows women how to be nurturing; and the Crone aspect shows that respecting elders is important and focuses on wisdom, change, and transformation.
>
> (1989 [1979]: 104)

Given that Hughes's myth of the goddess of complete being underlies his entire writing, the moon goddess that Hughes is referring to here can be read as his own personal tribute to his partner and muse Sylvia Plath. Moreover, there are plenty of moon images in Plath's own poetry: Plath's *Ariel* has been described as the classical 'muse of poetry, the moon goddess' (Moses, 2007: 101).

Although Plath dedicated *Ariel* to her children, bearing in mind that she wrote it after Hughes had left her, in the cold winter of 1962, it would appear to be aimed at Hughes.[13] There are certainly plenty of symbols that suggest that the figure of Plath was behind Hughes's thoughts when he wrote the poems. In 'Moon-Bells' Hughes talks of 'Holy bells' and reminds us that 'Love and Thirst, they know no shame' (*CPC*: 104) whilst mischievous foxgloves on the moon 'keep to dark caves' (*CPC*: 99). Given that the foxglove, or digitalis, universally accepted for treatment of heart disease, is a plant that heals the heart, it can also be read here as a symbol of healing.[14] Together with the moon goddess, many of the animals that appear in Hughes's *Earth Moon* (1976) poems appear as mythic animals traditionally linked with moon-worship, like the beautiful white Moon-Bull who stands at night 'Licking the glossy side/Of his infinite bride' (*CPC*: 107).

Hughes's moon poems were a project that absorbed him for over 25 years and can be read as part of his healing quest. They were written as a way for Hughes to come to terms with his own loss and perhaps sense of guilt, which becomes apparent in works like 'Orpheus' or his confessional *Birthday Letters* (1998), and as a personal tribute to Plath. It should also be remembered that the moon is one of the most frequently recurring images that appears in his poetry for adults. One of the most beautiful poems that Hughes has written depicting the image of the moon is 'Full Moon and Little Frieda', first published in the *Observer*, 27 January, 1963, only weeks before Plath took her life. In the poem Hughes captures the atmosphere of a summer evening and the actions of Little Frieda and the cows going home, whilst the moon expresses the fragility and innocence of a child's mind:

> 'Moon!' you cry suddenly, 'Moon! Moon!'
> The moon has stepped back like an artist gazing amazed at a work
> That points at him amazed.
>
> (*Wodwo*, 1967: 182)

Not only can the poem be seen as the reconciliation between humans and nature but as the last stanza brings to mind, perhaps Hughes was also suggesting a way for the adult to reconnect with the child's sense of wonder.

Set in the realm of Hughes's moon goddess, Hughes's moon poems depict a complex and contradictory imaginative universe, a dark but beautiful dream world, accompanied in his first edition of *The Earth Owl and Other Moon People* (1963), by R. A. Brandt's equally eerie black and white illustrations, and later by Leonard Baskin's in *Moon-Whales and Other Moon Poems* (1976). The fact that Hughes edited a special edition in 1976 which he illustrated himself, underlines the importance that the moon creatures must have had in his own creative universe.

The moon can also be seen as a projection of the poet himself, searching in the darkness to find his own voice. Perhaps the clearest example of this can be seen in the title poem of the collection, 'Moon Whales', which reads as a metaphor for the unconscious mind, depicting the crossover of borders and showing how scientific laws and facts do not apply to Hughes's surreal dream world. Enormous whales swim slowly through the lava seas creating the lunar landscape, and plough slowly along the surface 'lifting the moon's skin like a muscle' until they leave 'a hole blasted in the moon's skin' (*CPC*: 69).

Hughes's shamanic quest also appears in the disembodiment of the moon-shadow beggars who wander 'crossing the frontier from dark to light' (*CPC*: 76) in search of a body to invade. The poet warns of the shadows that reach out with long elastic arms and 'screech, sob and suffer in a dreadful way' cautioning us to be 'resolute, pass them without delay' (*CPC*: 76). There is a certain moral quality in Hughes's words here, against giving in to fear, as he warns that those who pity them will be lost forever and will be 'nothing but a skinful of shadows' (*CPC*: 77) and concludes that 'It is a horrible state and nothing can heal it' (*CPC*: 77). It is interesting to remark here that the notion of healing was already on Hughes's mind at the time of writing, together with an underlying sense of grief from his own loss.

Many of Hughes's moon poems reveal elements of gothic fiction, offering a glimpse at the 'dark' side of human nature, as in 'Moon-Ravens' which follow the black moth that 'flits out of your mouth' (*CPC*: 93) or in 'The Moon-Mare' which 'Wild as a ghost/She is here, she is past/In her lunatic fury' (*CPC*: 99). One of the identifying features of the Gothic is the fascination with boundaries, shifting thresholds like those portrayed in his moon poems where different worlds meet. Unlike much of the patronising 'light verse, trivia and nonsense' (Perkins, 1982: 189) that made up much of the bulk of children's poetry written during the 1970s, Hughes did not believe in talking down to children. As Hughes himself admonishes in a letter to Lissa Paul, 'children don't demand anaesthetics. One can communicate with children in a simple & whole way – not because they're innocent, but because they're not yet defensive' (*LTH*: 482). So, through the darkness of his moon poems, the poet seeks to offer the young reader a chance to experience the thrill of terror, by feeding their imagination with a disturbing awareness of just how scary the world can be.

It would appear however, that in this world of mourning Hughes does not find redemption in his moon poems. In fact, at times, Hughes does appear to move outside the boundaries of children's poetry, challenging the reader. Understanding the subject matter of some of his darker moon poems would appear to be out of reach of many children, and even young adults, especially in poems that are closer in style and content to Hughes's work for adults, such as 'Moon Thorns', which describes the way roses grow on the moon, with curved horns that 'stab you to the bone./While you groan' (*CPC*: 93). The almost chilling violent imagery of the poem carries with it an underlying sexual undertone, whilst the description of the thorn drawing a 'drop of your blood' (*CPC*: 93) can be read as a metaphor for fertility, intensified in the final couplet with the reference to marriage of the rose with thorns 'The horned bloom given/At marriages in heaven' (*CPC*: 93).

Indeed, despite their apparent simplicity, many of Hughes's moon poems are embedded with layers of meaning. Not only do they explore the fears, desires, dreams and nightmares of our subconscious, but they appear to be concerned with themes of crossing boundaries, whether physical or spiritual, imagined or real. They read as a kind of shared experience breaking down the barriers and challenging both the adult and child reader to reconnect with their inner self. However, one of the dangers perhaps of crossing over these boundaries is that in contrast to what children may find funny, adult readers, especially critics, will often find simply silly, causing them to shy away and failing to understand the essential indirectness of children's poetry, and the subconscious healing quality that Hughes believed could be found within.

A collaborative project: Baskin's art and Hughes's poetry: from *Season Songs* to *Under the North Star*

Of all the artists Hughes worked with, the most productive and influential partnership was that of the American sculptor and artist Leonard Baskin. The affinity between both artists was based on a sense of inspiration and admiration. Despite the fact that Baskin's collaboration is unquestionably one of the most sustained relationships of Hughes's literary career, very few critics have analysed its importance in depth.[15] The work of Baskin, like that of Hughes, had the qualities and essentials to disturb, yet at the same time portrayed an extraordinary beauty. Hughes himself said it was a 'startling, sinister beauty' (*WP*: 89), one that at first 'may shock us' (*WP*: 89) but which in the end 'makes us cherish it, as some source of elixir' (*WP*: 89). This is evidently true of collaborative works in collections of Hughes's adult poetry such as *Crow* (1970), or *Cave Birds* (1978) which depict grotesquely twisted and deformed black and white birds. However, Baskin's later work in collaboration with Hughes is much richer in the flora and fauna of the natural world, and less apocalyptic, especially in the case of his children's work in collections like *Season Songs* (1976) and *Under the North Star* (1981).

Season Songs (1976)

Season Songs is a collection of 24 poems, with six for each season.[16] The American edition, with watercolours illustrated by Leonard Baskin, appeared in 1975, whilst Faber published the British edition in 1976. However, *Season Songs* should not be conceived as a collaboration with Baskin as such, but more as a response or reaction on behalf of the American artist's first impression of Devon.[17]

The book belongs to a more settled period in Hughes's later life, which he spent writing and farming, and was dedicated to Carol Orchard, whom Hughes married in 1970. Throughout his life Hughes's close connections with farming can be seen at multiple levels showing the importance it held for the poet. From his early days as a young boy in Yorkshire and his wanderings across the farmlands of his native Mytholmroyd and later Mexborough; in his acquisition in 1973 of Moortown Farm in North Devon, where he would raise cattle and sheep with his wife and her father, Jack Orchard; to his later involvement with Michael Morpurgo from 1976 onwards in the Farms for City Children project. Hughes's career as a farmer coincided with one of the most prolific periods in Hughes's writing career, both for children and adults.

Both *Season Songs* (1976) and *Moortown* (1979) read as direct products of Hughes's farming venture and depict a new tone in Hughes's work: one of acceptance, followed by that of the celebration of life (and to a point, death) portrayed in the cyclical patterns of nature. In the case of *Moortown* (1979) the poems began as journal notes, recording Hughes's experience as a farmer. In contrast to his previous poetry for children, *Season Songs* (1976) would appear to have been written, under Hughes's farmer's hand, with a deeper sense of 'affection' or 'wonder', and has therefore a far more redemptive aspect to it. As Hughes wrote in a letter to Keith Sagar he 'deliberately made *Season Songs* upbeat, to buck me up' (2012: 56) – possibly after coming to terms with the fact that his brother Gerald, who had moved to Australia, was not coming back.[18] Unlike his previous children's works it also enjoyed an enthusiastic response from critics. Cushman praised it as being one of Hughes's 'most accessible and most obviously pleasing books' (1983: 246) and Sagar claimed that 'If *Crow* had been his *Songs of Experience*, *Season Songs* was to be his *Songs of Innocence*' (1978: 159).

The opening poem 'A March Calf' is a good example of how Hughes depicts the life/death cycle: 'Right from the start he is dressed in his best – his blacks and his whites./... Standing in dunged straw' (*SS*: 13). By means of anthropomorphism, the visual light tone is radically different from Hughes's other much cruder animal poetry for adults. However, the upbeat tone by the fifth stanza suggests that although the poem is not about death as such, it alludes to the calf's natural instincts and its inevitable tragic outcome: 'Ready for the worst, shut up in his hopeful religion' (*SS*: 13). By anthropomorphising the calf, Hughes ascribes the animal feelings such as joy and pain. Yet, the calf's happy unawareness of life is not of importance to the 'Hungry people [who] are getting hungrier' nor to the 'Butchers developing expertise and markets' (*SS*: 14). Far from underlying the ignorance of a calf that is bound to face its death at the slaughterhouse, Hughes is pointing here to the fact that death is an inevitable part of the cycle of life, especially in the animal kingdom.

Certainly the bucolic image of farming that most readers have remains a long way from the crude realm of truth and honesty that Hughes portrays. As David Whitley comments 'few of us now have any detailed sense of what actually happens on farms. It is an area closed off from our experience' (2013: 97). Despite its more upbeat tone, *Season Songs* (1976) should not be seen as being sentimentalist or indeed pastoral. Indeed, Hughes's own unsentimental poetic voice appears here as being much closer to his early adult animal poetry rather than to his earlier children's poems. Far from offering an innocent vision of nature, Hughes focuses on painful and delicate subjects such as the violence of fox and pheasant hunting in 'The Seven Sorrows' or in 'The Stag' where he tells the moving account of the suffering of a stag being hunted to death, to the point that 'his heart became just a club beating his ribs and his own hooves/Shouted with hounds' voices' (*SS*: 57).

The dualism of life/death is present throughout the collection, as in 'Sheep' accompanied by Baskin's sinister black and white drawing. The poem is narrated through the eye of the poet/shaman, who acts as a simple observer of nature, and bluntly informs the reader of how a mother sheep mourns her 'vanished lamb' (*SS*: 28), concluding that 'Death was more interesting to him.... So he died, with the yellow birth-mucous/still in his cardigan' (*SS*: 29).

As in many of his animal poems Hughes's anthropomorphism crosses over between the human/animal world, evoking the poet's own comments in *Poetry in the Making* (1967) on how to carefully observe and capture the spirit of a poem: to 'Imagine what you are writing about' (*PM*: 18) by focusing our whole attention and seeking the illuminating detail in order to capture the animal in words.

Both 'Swifts' and 'A Cranefly in September' are poems that speak to all our senses, bluntly at times, evoking with beauty the reality of both life and inevitably death. In 'Swifts' the poet describes his relief on seeing the birds return after migrating. The poem also voices Hughes's own ecological concern for maintaining the balance of the web of life. As he points out 'They've made it again,/Which means the globe's still working' (*SS*: 27). The rhythm of the poem accompanies the bursts of energy and speed of the bird's flight. The poem closes with the poet nursing an injured fledgling 'my little Apollo' (*SS*: 28) which he has carefully wrapped in a scarf but is unable to save it from its suffering and inevitable death.

In 'A Cranefly in September' Hughes uses imagery from the realm of myth and folktale, to describe how a cranefly, 'like a fairytale hero' is going through its slow death, whilst the poet, who is depicted as a 'giant who knows she cannot be helped in any way' (*SS*: 53), remains a helpless onlooker.

Unlike his previous children's poems depicting far away imaginative creatures and surreal universes, these are poems that focus on farm animals or local, domestic animals, remaining in reach of the reader. Set in the recognisable immediate landscape of farming Devon, Hughes is simply describing the natural world around him, the animal within its environment. *Season Songs* (1976) marks a turning point in Hughes's poetry, depicting 'a new relationship between the human and animal worlds' (Roberts, 2006: 124).

Hughes's perception of nature is that of being part of the web of life, so that the poet becomes 'a shepherd of being' as Robinson puts it (1989).[19] His perception of the animal world is also far more *real*: that is the creatures of the poems appear here as being very much present and tangible. Throughout the collection Hughes appears to be depicting a specific type of anthropomorphism, one that could be defined more accurately as *biophilic anthropomorphism*, in a more holistic sense drawing on E. O. Wilson's biophilia hypothesis (1984), and the underlying subconscious healing properties that arise from what he describes as a kind of inborn affinity, or innate human dependence on nature embedded in our genes. The poet's powerful understanding of animal life and the fact that he portrayed animals not as objects but as subjects also helps to instil a sense of sensibility and empathy in the young towards animals and the natural world. As Lawrence Buell has suggested, empathy with nature is key to the development of an environmental ethic: 'Both the awakening of obligation to become a reinhabitor and the awakening of a sense of environmental determinism require at some point reconceiving the human relation to the nonhuman, and the ethical borderline between these' (2001: 170). Hughes's collection not only reads as a celebration of nature, but as a word of warning about the fragility of our ecosystem. However, as Underhill suggests, the real value of Hughes's farming poems is not the fact that they illuminate a healing of the rift between humans and nature but that the poems depict 'a vivid reality to that concentrated specialization of human activity which farming increasingly is' (1992: 88).

Another theme that appears in the collection is the turning of seasons, as reflected in the titles of some of the poems. Whilst Spring begins with 'A March Calf' and shows 'The River in March', 'Spring Nature Notes: Parts I to 6' and ends with 'Deceptions', Summer shows the 'Swifts', together with more playful titles such as 'Work and Play', 'The Harvest Moon' or 'The Golden Boy'. Autumn is depicted as a cruel season, but contains some of the collection's best poems, including 'The Seven Sorrows', 'A Cranefly in September' and 'The Stag'. Winter is a picture of oppositions, as in 'The Warm and the Cold'.

Whilst the majority of the poems depict the realistic local landscape of the farmland, there are two poems that stand out as being distinctly different both in form as well as in content from the rest of the collection. One of them is the opening poem to Autumn, 'Leaves', written as a clever parody of the popular English nursery rhyme 'Who Killed Cock Robin?'.[20] In it the sacrificial death of the robin is reacted: 'Who's killed the leaves?/Me, says the apple, I've killed them all' (*SS*: 43). In Hughes's version instead of the thrush, it is the tractor's throttle which sings the psalm: 'Who'll sing a psalm?/Me, says the tractor, with my gear grinding glottal' (*SS*: 44). It ends with the winter robin singing in the bare branches: 'Who'll toll the bell?/Me, says the robin, my song in October/ Will tell the still gardens the leaves are over./I'll toll the bell' (*SS*: 44).

Rather than being read as a playful parody, Hughes's version of the rhyme can also be read as a wake-up call, bringing attention to a far more serious topic, evoking Carson's *Silent Spring* (1962), in which she warned us of the unforeseen consequences of pesticides, and sings the ballad, so to speak, lamenting the effects of DDT on the robin population. Hughes's own growing preoccupation in the 1970s with environmental issues such as the use of pesticides and their effect on the environment, the effects of pollutants on local rivers, together with factory farming led to his own local environmental activism. The complexity of these concerns underlie many of Hughes's farming poems, both in *Season Songs* (1976) and *Moortown* (1979) which express similar tensions whilst often questioning modern farming methods.

The other poem that stands out in *Season Songs* (1976) is 'The Golden Boy', which draws strongly on myth and folktale, in the form of the well-known English folksong 'John Barleycorn'. Hughes was familiar with James Frazer's *The Golden Bough* (1890), in which Frazer cites John Barleycorn as proof of the existence of a Pagan cult in England that worshipped a god of vegetation who was sacrificed to bring fertility to the fields. The link between myth and ritual ceremony is explored by Hughes as a story of celebrating the life/death cycle. When introducing the poem on the radio, Hughes himself commented that 'Growing grass and making hay of it ... is one of the deepest ceremonial satisfactions in farming, and every farmer feels it, no matter how mechanized or exhausted he might be' (Sagar, 1978: 243). Taking on the form of a prayer-like chant, the poem tells the story of a golden boy who was 'Buried in the dirt' but nurtured by 'the Lord's mother' until 'he began to live/And to thrive on her love' (*SS*: 39). It draws an apparent ironical equivalence between the harvesting of wheat and Christian sacrifice, in which the farmers are 'murderers', who bury the seed, and then reap the harvest, beating 'his bones' (*SS*: 40) and grounding 'him to powder' (*SS*: 40), then baking 'him in ovens' (*SS*: 40) only to slice 'him on tables' (*SS*: 40).

Critics have shown somewhat mixed opinions of both the poem and the collection itself. In the words of Sagar, the poem is 'an astonishingly fresh and vivid reworking of the John Barleycorn story' (1978: 161). However, according to John Stokes,

> it is difficult to know what to make of the irony when you do not know what you are supposed to make of the myth. Or rather what a child might make of them, for a child is even more likely to be bewildered by both.
> (1976: 256)

What remains clear throughout the collection is that these are poems written with 'affection' and 'within hearing' of both children and adults. More importantly, they enact Carson's sense of wonder in their unsentimental treatment of the truth of nature, depicting the cycles of life and death in all its splendour along with all its brutality offering a complete vision of the universe and portraying a sense of wholeness. The collection ends with 'The Warm and The Cold', which despite its simple appearance reveals much more on closer observation. The repeated form and regular rhyme scheme of the first three stanzas create a sense of balance and harmony. Despite the cold winter, Hughes depicts the warmth of the animals with similes such as 'the badger in its bedding/Like a loaf in the oven' (*SS*: 74), showing that, unlike humans, nature can adapt to the circumstances and be content with their lives. This contrasts with the double imagery of human/animal, in which the farmers, tucked up in their warm beds, are unable to put their minds to rest, and 'Turn in their sleep/Like oxen on spits' (*SS*: 75). Talking about the poem Hughes speaks of this sense of inhabitation of both the animal world outside alongside that of humans inside and comments on how animals endure the cold winter and how they 'must be in a sort of state of bliss to be able to go through what they go through and tolerate it' (*The English Programme*, 1998).

It is in poems like these, by evoking the relationships and boundaries of the human and animal worlds alongside each other, that Hughes puts forth his vision of the world as a whole and depicts the importance of finding a balance and reconnecting with the forces of nature and our inner universe. The real value of being attuned to nature is to see what lies before us, although sadly it would appear that most humans, like the farmers in the poem, are unable to perceive what lies before them, and remain unaware of the consequences of their actions.

Under the North Star *(1981)*

Out of all the collaborations between the poet and artist, perhaps the most successful was his children's work *Under the North Star*, published by Faber in April 1981. Despite the fact that it did not begin as a deliberate collaboration it proved to be much closer than any other before. Hughes and his second wife Carol, had gone to visit the Baskins in Maine in the autumn of 1978. Whilst there, both couples went on a trip to the North Woods, in Baxter State Park. During the visit Hughes kept a small black note book where he would put in writing impressions and drafts of poems. The book began as a birthday book for Leonard's daughter, Lucretia. Accompanied by Baskin's watercolours, these were more intense and, completely different from the illustrations in *Season Songs* (1976). As Hughes wrote, 'Several of them are small masterpieces' (Sagar, 2012: 91). Whilst *Season Songs* (1976) was not considered as a shared collaboration, *Under the North Star* (1981) was conceived as a joint project. What began as two or three little poems about birds in the North Circle ('The Loon', 'The Snowy Owl' and 'Woodpecker') was to develop into an internal collaboration between the poet and the artist, after Baskin suggested expanding it and making it into a children's book. The feeling of autumn pervades the whole book, which is set in true wilderness in the landscape within the Arctic Circle evoking the animals and birds that live under the North Star there. The final project proved to be very much a work of creative art.

Hughes and Baskin's 'affinity' as Baskin calls it in Noel Chanan's documentary *The Artist and the Poet: Leonard Baskin & Ted Hughes in Conversation 1983* (2009), developed into 'a relationship of presence' rather than 'a relationship of influence'.[21] Talking about the nature of the collaboration between the artist and the poet and in relation to the symbiosis between text and image, Baskin explains how he approaches the text, not attempting a visual counterpart, but instead as an extension of the text and its meaning: 'That's what great illustration is: it expands and extends one's understanding, rather than provide a visual nomenclature (so to say) of what's happening in the words' (Chanan, 2009).

It is precisely that idea of extending the text and giving insight, especially when approaching subjects such as animals that readers may not be entirely familiar with, that appears in many of Baskin's collaborations illustrating

Hughes's children's works. *Under the North Star* (1981) achieves this idea of extending the poem beyond the written text so that, as Peter Cook and Mick Gowar have pointed out, 'It is impossible to tell which came first, poem or painting, such is the depth of correspondence between them' (2011: 2).

Like in *Season Songs* (1976), Hughes acts as a humble observer of nature. There is a clear sense of interconnectedness between the animal/non-animal world present throughout the entire collection. In the opening poem 'Amulet'[22] the world and the forest become interconnected, so that the external world becomes part of the creature's inner self, each embedded in each other. In the poem Hughes attempts to control the power of the wolf by constructing a poetic charm, and so the wolf's fang becomes the amulet in the case of the poem. The Wolf appears with a capital letter each time it is used and is the central image. The poem is circular in form and rhythm, with each line referring in some way to a part of the creature, and each image building on the previous. Moving through the verses, the poet imagines how each object might connect with the next until the circle which began with the wolf's fang in the first line is completed and closed in the last line of the poem with the wolf's fang. In this way the rhythm of the poem and the circular narrative progresses as a magical incarnation, like the shaman's drum, and ensures that the wolf's power is contained within the verses.

Given the cyclical process that takes place throughout the text itself, the poem can also read as what John Elder has described in the field of ecopoetry as a kind of 'ecosystem' in itself (1996: 39), in terms of the interrelation and illumination of the amulet's charm throughout the poem. The images that are associated with the wolf follow a progression, leading us through each connection on a symbolic and aesthetic journey, almost like a poetic chant representing the wolf's nature and allowing the reader to enter the encircling magic of the powerful amulet, under the spell of the poet/shaman.

Once again, critics appear to be divided. Whilst Cushman praises Hughes's simplicity, and states that *Under the North Star* 'could be called Hughes's *Birds, Beasts and Flowers*' (1983: 251), Anthony Wilson erroneously claims that its failing is that the reader is not given enough time to familiarise themselves with the animals 'before the overruling metaphor of the sheer inaccessibility and inhospitality of the world they live in takes over. This may well be Hughes's point, but it does not make for reading that children would find engaging' (2001: 83).

Perhaps it is Wilson who is missing Hughes's point here, for Hughes does create a metaphor of inhospitality, but intentionally, for humans at least, in his depiction of the wilderness of the Arctic Circle which acts as an ecosystem, like the poem itself. Although Hughes's landscape is one that is harsh, it is also beautiful; the stark white Arctic Circle is evoked as a mystical land, where its creatures are both totemic and real. Inside this magical, but realistic landscape the animals that live there are brought alive, through the words of the poet, and from the rich, luminous brushstrokes of the artist's imagination, looking out disconcertingly at the reader. The pictures focus on the almost tangible faces of the animals. From the deep stare of the lynx, to the penetrating eyes of the snowy owl, or the grizzly bear's gaze which occupies the entire page, to the sideways glance of the oversized moose who is too large to fit on the page or that of the eagle on the last page.

Despite its format and appearance as a children's picture book – a series of double page spreads with Baskin's large A4 size picture on one side accompanied by Hughes's large print on the other – the poems remain far from the realm of classic children's picture books. Although decidedly less sophisticated, and mostly written in free verse, like *Season Songs* (1976), most of the poems can be read as crossovers between the borders of adult/children's poetry. In many cases Hughes successfully takes the reader beyond the image of the animal into the landscape where he is writing. Such is the case of 'The Snow-Shoe Hare' who is 'his own sudden blizzard' (*UNS*: 24) or 'The Arctic Fox' who sleeps in 'the snow-lit land' and as the last stanza beautifully describes: 'When day wakes/Sun will not find/What night hardly noticed' (*UNS*: 38).

There is, however, another aspect underlying the poems that seems to have gone unnoticed by critics, Hughes's deliberate play on religion, or rather the idea of sacredness that can be found in his Swiftian idea of divinity, and the explicit references that appear throughout the collection towards Hughes's spiritual idea of the divine in all living things. Examples of these can be found in 'The Snowy Owl' when Hughes proclaims 'Let us all rejoice in the Hare!' (*UNS*: 16); and again when he closes the poem with the metamorphosis and exclamation that 'The Hare has become an angel!' (*UNS*: 16). There are more unmistakable religious references, as in poems such as 'the Black Bear' where the bear condemns God for having created him 'God swore/I'd be Adam/... But I am in a cage' (*UNS*: 18).

In a similar way in 'Brooktrout' the poet describes how 'You can catch Brooktrout, a goggling gewgaw –/But never the flash God made/Drawing the river's blade' (*UNS*: 32). In 'The Osprey': 'God hides below/In his shadow' (*UNS*: 34), whilst in 'Puma' we learn that 'God put the Cougar on the Mountain/To be the organist/Of the cathedral-shaped echoes' (*UNS*: 39). Certainly in all these religious references towards the non-human Hughes portrays the creatures of the world as divine forms. Likewise, in 'Mooses' the goofy Moose appears to be lost, out of place, unable to find the world he once belonged to: 'Where is the world?' he groans, 'O my lost world!' (*UNS*: 36).

Perhaps the most remarkable poem and one of the most successful in portraying the union between 'the other' and its environment, as well as Hughes's effort to speak for nature, is 'A Lynx'. Moreover, it seems to be delivered with what Hughes called 'affection'. The symbiosis between Baskin's illustration and the poem itself is outstanding. The lynx's red-brown fur stands out against a black background depicting the forest at night. The union between the beauty of the lynx and the mountains depicted in the first stanza contrasts with the poet's warning not to 'disturb this beast' (*UNS*: 22) that opens the second stanza. Acting as a steward for nature, the poet's caution is directed to those who observe, who listen to the poem, and who, as the poet reminds us, do not form part of the landscape. Hughes seems to be reminding us that if we disturb nature, if we attempt to disrupt the harmony of the natural world not only are we altering its natural balance, but we are committing a crime against nature.

All the creatures that live under the North Star live in harmony with the forces and energies of the universe. The images of bleakness and violence that appeared in *Season Songs* (1976) are more balanced by the images of harmony depicted in *Under the North Star* (1981), such as those conveyed by the static images of the Wolverine, the Snowy Owl or the Snow-Shoe Hare. It is this sense of capturing the creature and observing its behaviour in the narrative environment created for it that the poet portrays so beautifully, allowing the reader to enter the internal and external world of his own creative poetic universe, a world written for children, but as Hughes reminds us 'which adults can overhear' (*LTH*: 482).

An aspect that stands out particularly in 'A Lynx' but which can also be found in other animal poems is the way in which Hughes invites the young reader into the negotiation of responsibility with 'the other'. The poet gives representation to the non-human voice of 'the other', or 'the *another*', as an entity with its own voice.[23] In many of his animal poems, and especially in works such as *Season Songs* (1976), or *Under the North Star* (1981), 'the other' is not treated as a mere object, but considered instead as a subject, as part of a community.

It is this interconnection and Hughes's desire to understand the more-than-human world around us, especially the poet's depiction of the animal within the space of its environment, that allows the reader to discover that 'sense of wonder', to the point that the narrative environment of his poetry can be read by both the child and adult reader alike.

Hughes and Lloyd: from *What is the Truth?* to *The Cat and the Cuckoo* and *The Mermaid's Purse*

Despite having been an ongoing commitment for nearly 25 years, Hughes's partnership with Reg Lloyd is less well-known and documented. The poet and artist had developed a working friendship as farmers and both shared a passion for nature and a reciprocal sense of respect for each other's work. Their first collaboration dates back to the 1970s with Lloyd's designs for broadsides for four 'Crow' poems and continued throughout the 1980s and 1990s with his children's books *What is the Truth?* (1984), *The Cat and the Cuckoo* (1987) and *The Mermaid's Purse* (1993) as well as and Lloyd's personal selection of Hughes's poems, *Earth Dances* (1994),[24] together with plans for other collaborations which sadly never took place.

What is the Truth?: A Farmyard Fable for the Young *(1984)*

For most readers and critics alike, *What is the Truth?* is one of Hughes's most successful books. With it Hughes won his third *Signal* Poetry Award in the same year as he was appointed as Poet Laureate.[25] Written as *A Farmyard Fable for the Young*, it reads as a book-length story in verse. Morag Styles called it 'Hughes's masterpiece for children' (1998: 254), whilst Neil Philip acknowledges that although it is 'inevitably uneven' it is 'packed with genuine poetry: harsh, funny, urgent, potent. The whole book … has a marvellous quality of improvised speech, jingle and song' (2003: 197). The unevenness that Philip (2003) points to, the fact that it is halfway between story and verse as well as being a picture book, would appear to be one of the book's major drawbacks. The overlapping of prose sections, which act as a framework to the poetry, is at times not particularly well demarcated, so that, as Thomas West says, the main problem 'is the rhetoric of the different elements in the story, the relation between them or, more precisely, how the reader understands the various elements' (1985: 111).[26]

Dedicated to 'Clare and Michael', the book was written at the suggestion of Hughes's friends, Michael and Clare Morpurgo, who ran Farms for City Children in Devon. Hughes became an important figure in Morpurgo's life, acting as a source of inspiration to the writer, and showed genuine sympathy towards the project. He became their first president, whilst actively involved helping with fundraising and reading to the children. Hughes himself wrote about the genesis of the book in a letter to Nick Gammage, in which he claimed that he

> floated the notion of a fable about the true divinity of creatures and man's inability to see that truth … Mainly I wanted to write pieces that would make it easier (even easier?) for readers – urban readers – to like these creatures. Or to share my liking.
>
> (*LTH*: 571–572)

Drawing again on the idea of 'divinity', Hughes links the notion to that of empathy, a key concept underlying Hughes's anthropomorphism and which from an ecocritical viewpoint is not only considered as being necessary in order to create a sense of environmental imagination towards nature in children from an early age, but one which is also prior to that of responsibility in the reader.

Hughes's sense of 'divinity' seems to have a different inspiration here from that of works like *Crow* (1970), and is closer to that of paganism. The poet uses the Bible here as a narrative framework, one which is familiar to the reader and one within which he can explore his own healing quest and the human/non-human relationship that underlies the tale. In a draft found in the archives at the British Library, Hughes himself offers a more insightful explanation to the idea behind his fable, explaining that as president he was asked to write something about the animals on the farm:

> If I could direct their attention, and connect what they could observe with a playful verbal approach, giving many examples of a provisional attempt to capture, in words, something of the reality that could be glimpsed, I thought it might be worth doing – and might even encourage them to make their own playful, provisional attempts.[27]

Hughes's ideas about connecting words to experience are close to those in *Poetry in the Making* (1967). There is also an allusion to Carson's 'sense of wonder' underlying the whole text, especially in Hughes's awareness of how young children responded to such an experience. As Hughes, like Carson, was aware, it is in those early years that the child's environmental awareness is shaped, especially through positive encounters and hands-on experience like that of Farms for City Children, which encourage children to develop their sense of biophilia, their need for nature and knowledge and appreciation of the natural world, leading towards the development of an environmental ethic. Sadly, as David Whitley has pointed out 'Children's loss of the ability to recognize and name – let alone understand and relate to – the real animals that share their local environment has become one of the most profound indices of our current crisis with nature' (2013: 106).

The narrative frame to *What is the Truth?* is a story about God and his Son, both of which appear capitalised in the text. The text opens with God's Son wanting to visit earth to learn from mankind. God tries to warn him that 'Mankind cannot teach you anything. Mankind thinks it knows everything. It knows everything but the Truth' (*WT*: 9). But God's Son insists so God decides to give him a moral lesson and the two come down to earth in the middle of the night and begin to question the spirits of the farmer, his wife and children and other village characters. God asks them about the nature of the creatures they know and love. However, God is unimpressed and explains that none of the spirits have spoken the 'truth'. To which his son asks him 'What is the Truth?' (*WT*: 121).

Throughout the moral tale it becomes apparent that the 'truth' is not a scientific fact or an abstract reality. This is an idea that Hughes had already played with in his children's plays, as in the case of Dully and the master in 'The Tiger's Bones'. Thus, the story can also be read as part of Hughes's own shamanic healing quest as an attempt to explain his own quest for the 'truth'. The poems seek to establish the truth about the nature of God's creatures: animals, birds and insects, and their relationship to human beings. Whilst the people of the village remain blind to the truth and fail to be spiritually awakened, their versions of the truth are only partial truths, as revealed in the final passage of the book in the words of God: the truth that everything that lives is holy and that 'each of these things is Me. It is. It is. That is the Truth' (*WT*: 121), so God and the animals are one and the same. It is by connecting the divine with the ordinary and the human with the animal, that Hughes allows the reader to discover and realise the reality of our world, and to show that the real 'truth' resides in what lies before us: in nature itself, in every living thing.

Hughes's own 'poetic truth' is of course similar to that of Robert Graves.[28] The notion of truth that both poets use is close to the biblical sense of truth and underlies the idea of sacredness in all living things. In this context the book can be read as a hymn to nature, a celebration of the web of life to which we are all connected. There are also similarities between Hughes's poetic 'truth', revealed here in the words of God, and the Blakean sense of truth, as reiterated in *Songs of Innocence and Experience* (1789), that every living thing is holy.

Critics such as Skea (2012) and Sagar (1983) have interpreted Hughes's poetry as being 'fundamentally akin' (Walder, 1987: 45) to that of Blake. However, more than suggesting a religious, or spiritual connection, it seems that, in true Swiftian style, Hughes grants the status of 'divine' to the animals here, who 'are their [the villagers] toys' (*WT*: 99) and not the humans. In fact, whilst the animals are considered as sacred, it is the human beings that are placed under examination and judged because they are only able to offer what Blake called a 'single vision',[29] in other words they can only see the empirical world that lies in front of them. Unlike adults, children are alive to the world of imagination. Hughes firmly believed that if adults were able to awaken their senses and reconnect with that 'sense of wonder', they could retrieve the vitality and become aware of the sense of active imagination.

Another fundamental element of the collection is, of course, Reg Lloyd's illustrations. Despite the fact that Hughes firmly believed that the image took the viewer's attention away from the text, it was originally Faber and Faber who had insisted that being a children's book it needed illustrations and had forced them onto Hughes – who finally agreed. However, it was not Lloyd, nor Hughes, but Hughes's wife Carol, who suggested that Lloyd should read the manuscript of *What is the Truth?* with a view to illustrating it. After submitting a few monotypes to Faber he was commissioned to illustrate the book working from a list of animals and birds that Hughes had given him and became the first of three collaborations for his children's books *What is the the Truth?* (1984), *The Cat and the Cuckoo* (1987) and *The Mermaid's Purse* (1993).

In a retrospective exhibition of Lloyd's work, and referring to *What is the Truth?*, Hughes wrote that:

> it struck me at the time, there is hardly a plate in the book without its heavenly light – a moon or sun. Even the worms, in one of the best, seem to be worshipping the rising sun. One way or another, he has made each of these creatures a sacred figure.
>
> (1996: vi–vii)

The motif of the moon or sun accompanies nearly all of the drawings, so that each of the figures appears as 'sacred'. The book reveals an extensive and challenging synthesis of word and image with Lloyd's monotypes on almost every page. Lloyd also particularly enjoyed the technical challenge of designing the layout of the book which captures both the essence of the text and the spirit of the animal. Each animal or bird is depicted with a fresh and powerful underlying sense of life to it.

Lloyd's 116 illustrations were first drawn on stained glass and then as a monotype featuring animals of different sizes and habitats: both farm animals and animals from the wild. Rich in life, they beautifully accompany the text, whilst their gaze captures the reader, each one lyrical and symbolic in meaning, so that the animals are portrayed both in words and image, and depicted with 'affection'.

Many of the poems in the collection reveal an extraordinary simplicity and effortlessness about their language and form. One of the best examples of how these poems are 'simple' but 'rich in life' and how Hughes captures the essence of real animals can be seen in 'Roger the sheepdog', who lies and 'wheezes at his ease' and 'hogs the fire' (*WT*: 107). The long vowel sounds in the first line set the calm and lazy tone for the poem, whilst the similes used by Hughes bring the reader inside the text and evoke the image of an older lazy dog, depicting to perfection how the dog behaves: only waking from his sleep to eat and scratch his fleas.

What is the Truth? also allowed Hughes the opportunity to incorporate poems from other collections, and as a cross over between adult/child readers. One of Hughes's most celebrated 'adult' poems 'A Solstice' is an interesting case.[30] Having first appeared in 1978, and written as an attempt to recapture Hughes's childhood days through a hunting expedition with his brother Gerald, it appears here for the first time in a children's book, a fact that appears to have gone unnoticed by Sagar and other scholars. Furthermore, in this context it undergoes a change in narrative tense from the present in the original to the past, as well as other minor alterations: 'Four days to Christmas' becomes 'The Day Before Christmas' (*WT*: 112). The fox 'a slender figurine, dark and witchy' (*WT*: 116) is also 'An elegant gentleman, beautifully dressed' (*WT*: 117) recalling Beatrix Potter's emblematic anthropomorphised fox figure 'an elegantly dressed gentleman' (1989: 22) from *The Tale of Jemima Puddle-Duck* (1908), which Hughes would have been familiar with and had probably read to his children. In the context, and despite the sinister tone of the hunt, the poem reads as a somewhat unlikely children's poem.

The book also recalls the format of the medieval bestiaries, drawing on the language of the illustrations, with the exception that whilst the former attempted to be moralising and show the differences between man and beast in the case of Hughes and Lloyd, he attempts to bring the two together. At the end of the book, *What is the Truth?* turns itself into a true picture book, showing pictures of different animals with captions reading 'I AM THE RAT', 'I AM THE CAT' until the last page displaying nine pictures with 'I AM ALL OF THESE THINGS' written underneath in capitals.

Hughes also makes an interesting case in the tale against keeping wild animals in captivity. In the thought-provoking poem 'Bess my badger' the farmer's son tells how a badger grew up 'In a petshop in Leicester. Moony mask/ Behind mesh' and describes her 'Trudging to-fro, to-fro, in her tight cage/... She was/Learning to be a prisoner' (*WT*: 13). Hughes disrupts the human–animal hierarchy of dominance here since it is the badger who perfects her behaviour as the perfect prisoner through captivity and then takes over the space of the barn redefining human/animal space and inviting the reader into a reverse perspective so that at the end of the poem 'she's our houseproud lodger, deepening her rooms./Or are we her lodgers?' (*WT*: 14). The rhythm of the poem mimics the movement of the badger stalking its cage. When the badger is let free, and returned to the wild Hughes places the badger, as 'the true king of this land'. Another example of the way in which Hughes plays with human/animal relations and identities can be found in 'The Fox'. In the poem the farmer finds an unconventional parallel in the figure of the fox where 'The Fox is a jolly famer and we farm the same land' (*WT*: 34). He sees his animal equivalent as 'a hard-working farmer, with a farmer's hard hand' (*WT*: 34).

It is these kind of challenges that Hughes plays with by crossing the boundaries between human/animal. By recognising animals as beings of significance, the capacity for human compassion grows, encouraging respect and raising ecological awareness.

In this period of his poetry although his children's poems are written in a simpler style than his adult collections, the themes and subject matter are similar. As Cushman suggests, it is during this phase of Hughes's development as a poet that 'the adult and children's poems have drawn closer and closer together' (1983: 254). Given that Hughes did not write down to children, like his other animal poems these are also unsentimental and sincere, written with the insight of a poet and farmer who has worked with animals, as in the case of the shepherd who tells God about 'the Problem about Lambs', sparing no details in his description of the ulcerous infection that affects sheep 'After that comes Orf. Known as Lewer./Ulcers of the nose, of the lips, of the eyes, of the toes' (*WT*: 36).

What is the Truth? reveals a structural redemptive process in the sense that the book can be read as a healing vehicle for young readers and of course their parents. By examining each of the animals' place in the ecosystem Hughes calls for an ethical relationship with the natural world and encourages the reader to learn about the habits of animals through indirect observation.

It is not incidental that Hughes opens the collection with a poem about a 'profane' truth: the death of partridges and how they become the farmer's dinner, and closes the collection with a 'sacred' poem about the wonder of two lobworms in love: 'two bare creatures, stark naked, were they/full length in the mud, at the dawn of the day' (*WT*: 119). The poet describes them with the sensuousness of two passionate lovers who 'twisted together like two loving tongues/And they had not a care for the world and its wrongs' (*WT*: 119). In many ways the poem recalls Hughes's 'Bride and Groom Lie Hidden for Three Days', from *Cave Birds* (1978), originally intended to provide the culmination of Crow's efforts to understand the world in which he finds himself: 'Like two gods of mud/Sprawling in the dirt, but with infinite care/They bring each other to perfection' (CB: 56). Like 'Bride and Groom Lie Hidden for Three Days' in *What is the Truth?* the 'two bare creatures' portray a redemptive image of resurrection and regeneration in attempting to transform mud into God. In 'Bride and Groom Lie Hidden for Three Days', according to Gifford and Roberts, 'The "marriage" between the two is the achievement of wholeness' so that 'the lovers "bring each other to perfection"' (1981: 222). This sense of wholeness and union is also achieved in the case of the lobworms, bringing to mind, of course, Hughes's myth of the goddess, and the creative/destructive laws of nature that govern his entire universe of poetry, so that the redemptive vision of healing that Hughes engages with throughout his poetry can be seen in the symbolic marriage of humans and nature in both poems.

Another wedding appears in the poem 'Bees', which was later added to the published version (*LTH*: 455). In it, the bees 'fall/On to their knees, and humbly head-down crawl/Into their crammed church' (*WT*: 58) and end in a peculiar religious act of blue-gluing where the priest bee 'in a shower of petals,/ Glues Bride and Groom together with honey' (*WT*: 58). Despite its more upbeat tone, the poem inevitably recalls Hughes's haunting 'Bee God' (from *Birthday Letters*, 1998) which examines his relationship with Plath and of course Plath's own bee poems.[31]

In the final passage of the book Hughes tells us how God's Son 'hesitated. It occurred to him, with a little shiver, that he was where he had wanted to be' (*WT*: 114), in the world portrayed with all its 'wrongs' (including man's alienation from nature). The real truth put forward is that of the individual nature of animals and their relationship with human beings. Moreover, despite its 'wrongs' the earth, with all the living things that inhabit it, appears to be the best place to be. In this way the duality of nature/culture is redeemed and united through the poems and their ending. The final poem can also be read as a song of redemption, written, like his other poetry from this period, for children but 'within hearing' of adults and conceived as part of Hughes's healing quest. Hughes firmly believed that poetry could heal us, by shocking the reader's imagination and putting us in touch with nature. In Hughes's words, it is:

> As if poetry were a biological healing process. It seizes on what is depressing and destructive, and lifts it into a realm where it becomes healing and energizing … And to reach that final mood of release and elation is the whole driving force of writing at all.
>
> (1978: n.p.)

Hughes's engagement with the notion of truth, that of the universal truth of sorts, about man, about human nature, that lies in his collection depicts a new relationship to the natural world, one that can be described as a sacramental union of animal and nature and, at the same time, one close to the Christian appeal to cherish and revere all living beings. In the local animals depicted in his farmyard poems he recognises the holiness of nature, as a way to restore the balance of power in the natural world. As an eco-poet, recognising the interdependent nature of the world, Hughes can be seen as acting here as a kind of 'green' Noah, one who speaks on behalf of nature, presenting us with his poetic vision of the 'truth': that is the revelation depicted in the balance between humankind and nature in the ecological web of life, where humans and animals are one and the same, and where the divine lives not in God, but in the creatures of the earth.

The Cat and the Cuckoo (1987)

Shortly after completing *What is the Truth?* (1984), Reg Lloyd suggested to Hughes to do another 'small richly-coloured books about familiar creatures' (Skea, 2012), and publish it themselves. The result was *The Cat and the Cuckoo* (1987):[32] a series of 28 separate colour illustrations of animals, painted by Lloyd each with a poem by Hughes printed below, published as a limited edition of 2,000 books.[33] An interesting feature of the book is its distinctive small format, which recalls Beatrix Potter's animal tales, especially made for children to hold.

The poems that appear here are certainly 'less complicated, less sophisticated, more juvenile' (*LTH*: 572) than those that Hughes wrote for *What is the Truth?* (1984). The conception of the child reader that Hughes addresses here is shaped by his imaginative and witty rhyme which is intended for a slighter younger audience. Hughes dedicated the collection to 'Daniel Hews [*sic*], Sebastian Clarke, Hugh and Toby Norton-Smith and for all the children who visit *Farms for City Children*'.

Like his previous books the themes focus on the natural world: the local animals and their natural habitat and the neighbouring English countryside. Hughes himself said that 'my plan was to compress each subject into a very brief but intensely musical form that would be above all easy to memorise' (Neill, 1995). It would appear that Hughes is referring here to the idea of compression in his verse: whilst his first books of poems contain much longer and denser verse (like *Meet My Folks*, 1961, or *The Earth Owl and Other Moon People*, 1963) or were written in poetic prose (*Nessie the Mannerless Monster*, 1964), his last books of poetry for children are written in a much shorter and compact verse and with a clear marked rhythm, making them more appealing to the young reader. Curiously, the painting on the cover of the book, and the first poem in the collection 'Cat', depicts, in fact, the Lloyd's family cat, Tim: 'You need your Cat./When you slump down/All tired and flat/With too much town' (*CPC*: 19).

Although the poem begins lyrically with a simple short rhyme pattern, Hughes reminds us that the domestic cat is still a wild cat, which belongs to the family of felines, whilst describing the therapeutic effects found in the companionship of a pet, 'And you'll be refreshed/Through the Cat on your lap/With a Leopard's yawn/And a Tiger's nap' (*CPC*: 19).

Many of the poems mimic both domestic and farm animal sounds and movements, as in 'Peacock', where 'A perfect Peacock on the lawn/Pranced proudly through his paces' (*CPC*: 33). Despite the minimalism of the poems and their word play, the underlying idea of the poet's vision of nature as something that humans have disturbed is still present throughout the collection, in which Hughes alerts the young reader to the excitement and pleasure that can be found in coming into contact with the natural world and its creatures.

In a letter to Nick Gammage in 1998, Hughes states that both *What is the Truth?* (1984) and *The Cat and the Cuckoo* (1987) were written as 'a fable about the true divinity of creatures and man's inability to see that truth' (*LTH*: 571).

Hughes's support of farms and his own career as a farmer give evidence to the creativity that lies in his animal poems and to his central function as a poet. Acknowledging the poet's sense of the divine, Gammage had written to Hughes and discerned, in *The Cat and the Cuckoo*, a tendency to equate religion with hypocrisy. Indeed there is a dark sense of religion underlying poems such as 'Thrush' where the thrush rejoices to the snails, 'God's here again!/Close your eyes for prayers/While I sing Amen' (*CPC*: 20) together with a more serious metaphysical meditation in poems such as 'Ram' or 'Crow'. 'Ram' tells of the sleepless nights of a Ram who can't sleep and instead of counting sheep wonders 'where was I before?' (*CPC*: 26); meanwhile in 'Crow', 'Nature to Nature/ Returns each creature' (*CPC*: 27) alluding to the sacredness of the natural world.

In reply to Gammage, Hughes admitted that: 'Maybe I automatically draw on religion – the ultimate field of hypocrisy – as an analogy for the dissembling, hoodwinking and general self-misrepresentation by which so much of the organic world survives and even makes its living' (*LTH*: 572). In his letter he also refers to 'the crisis of modern thought' (*LTH*: 573). This was, of course, what Hughes feared had caused human's alienation from nature and, what he considered as a consequence of 'man's biological misfortune', that is 'man's talent for creating religious and political systems which turn him against both inner and outer nature'.[34] Hughes's allusion to the problematic split between human/nature in Western thought also underlies that the irony of the human condition is that man is an unnatural creature in a natural world. In line with Wilson's sense of 'biophilia' and Hughes's own healing quest the poet seeks to reconnect Western culture correcting the human alienation from nature and our inner selves.

On the whole *The Cat and the Cuckoo* may be a collection of simple nature poems which depict local familiar animals, however the latent meaning of the context recall that a poem, as Hughes says, is 'an assembly of living parts moved by a single spirit' (*PM*: 17). Hughes called this notion of poetry being alive 'the goblin in the word' (*PM*: 18) one which 'the poet has to have under control' (*PM*: 18), and one which his poetry for the very young, as in *The Mermaid's Purse* (1993), appears to lend itself to particularly well, inviting the young reader to 'see' the subject, crossing over the barriers into the otherness of animals.

The Mermaid's Purse *(1993)*

The next book that the poet and artist collaborated on was *The Mermaid's Purse* (1993). The genesis of the project was similar to that of *The Cat and the Cuckoo* (1987): Hughes first wrote a list of possible subjects, following which Lloyd went off and did the illustrations whilst Hughes wrote the poems. That Hughes would devote an entire book to the sea makes sense. On the one hand it could be seen as a personal tribute to Rachel Carson's *Under the Sea-Wind* (1941), and *The Sea Around Us* (1951), which Hughes and Plath had both read and been greatly influenced by. On the other hand, in allusion to the title, 'The Little Mermaid' was one of Plath's favourite childhood tales.[35] In either case *The Mermaid's Purse* was to be Hughes's last book of poetry for children. By focusing on a part of the natural world that he had not written about before, Hughes found the poems more challenging to write as well. In a letter to Skea he said that, 'It's hard to write a quantity of poems about sea-creatures. Some of them are so cold and prickly for children'.[36]

The 28 poems that make up the collection depict the different creatures of the sea. The title makes reference to the egg cases of sharks, dolphins or skates which are frequently washed up on beaches, although in Hughes's imagination, a shark leaps from the mermaid's purse to cure her headache by biting off her head. Hughes's whimsical sense of humour may be somewhat difficult for critics to deal with, but would appeal here to the young child's imagination. Written in Hughes's distinctive voice and with its underlying appreciation of the natural world *The Mermaid's Purse* explores the mystery of the ocean. Like Carson, Hughes writes about the daily struggles of the different sea creatures, and the drama of survival between hunter and hunted in their daily life. As Carson revealed, 'in the sea nothing is lost. One dies, another lives, as the precious elements of life are passed on and on in endless chains' (2007 [1941]: 105). In the first poem of the collection 'Seal' Hughes uses empathy in his anthropomorphic depiction of the seal and in his use of first person. The poem tells of how the poet, through the seal's gaze, thought that 'It had lost its mother/In the sea' (*CPC*: 3). Other creatures include the sea-anemone in which 'many an enemy/Melts in my embrace' (*CPC*: 5), or a shell in which 'you may wash out the sand/But never the sound/Of the ghost of the sea/That is haunting me' (*CPC*: 11).

All the poems however have a central point that unites them, that of the symbolic figure of a wreck in which 'The sailors prayed to come to land' but which lies under the cliff as 'the sea's rare rust-flower' (*CPC*: 7). Lloyd himself wrote about the genesis of the wreck, 'Ted, Danny Weissbort and myself had walked to Hartland Cliffs where the ship "Johanna" had come ashore and was now well rusted. The twisted metal plates standing on end suggested some sort of growth' (Skea, 2012: n.p.).

The concept of the wreck can also be seen as illustrating the symbolic union between humans and the sea. The different creatures live close to the wreck, so whilst the gulls glance overhead dipping and diving above the sea spray, the cormorants have taken the bodies of those who have drowned, 'Drowned fishermen come back/As famished cormorants/With bare and freezing webby toes/Instead of boots and pants' (*CPC*: 5). Other creatures from underneath the ocean sing their sea-soaked poetry, from the stones on the beach, to the crabs in their pools, to the poem 'The Monster' which depicts a sea monster rising above the surface, strongly reminiscent of Nessie.

The Mermaid's Purse can also be read and understood as Hughes's desire to create awareness in the young reader for an ecosystem which human beings have been complacent about for centuries, and which judging from the state of our oceans today have still not yet been fully understood.[37] As Carson already suggested over half a century ago, 'Humans are not fully adapted to life in the oceans, and we do not see or grasp the full extent of our impacts beneath the seas' surface' given that 'Much of the damage we do to the oceans is *out of sight, out of mind*' (Sideris and Moore, 2008: 90). Like Carson, Hughes urges us to use all our senses and explore the natural world with our feelings and emotions. By using imagination and Carson's sense of wonder, Hughes recreates an imaginative marine world, allowing us to value and understand how the ocean's ecosystem functions, by means of creating empathy in the young reader.

So ultimately, Hughes's children's poetry reflects his way of thinking about human–animal relations and holds promise for healing the bonds between humans and nature. Drawing on the roots of ecopsychology, Hughes's poetry suggests that there is a synergistic relation between the human and non-human – one which begins in childhood but usually fades with adulthood, through our disengagement and alienation from nature.

As revealed, throughout his poetry, and as part of his overall project, Hughes's sense of nature becomes more holistic. Instead of writing about a physical nature that is untouched and separate from human being, his poems bridge the gap between the human/non-human allowing the reader to cross the boundaries between the human and 'more than human' world. Using Timothy Clark's definition of ecopoetry, Hughes's poems are 'conceived as a space of subjective redefinition and rediscovery through encounters with the non-human' (2011: 139). It is this interconnection and especially the poet's depiction of the animal within the space of its environment, that allows the reader to discover that 'sense of wonder', to the point that the narrative environment of his poetry can be read by both the child and adult reader alike.

Given its upbeat underlying tone and intended audience, or at least readership, it would appear that Hughes best articulates the apprehension of a sense of balance and interconnectedness necessary for humans in his so-called 'poetic truth' in the realm of his children's poetry, for this is where the connection between the human/animal is perhaps clearest and where the adult/child reader can also be reconnected. Hughes's quest of truth shares Blake's belief that imagination is the main source of energy, and one which leads to the concept of truth, in line with John Keats's assertion that 'What the imagination seizes as Beauty must be truth'.[38] Hughes addresses Blake's vision in *Poetry in the Making* (1967) when he says that 'as a poet, you have to make sure that all those parts over which you have control, the words and rhythms and images, are alive' (*PM*: 9). Hughes's poetry, however, does not compromise for the young and goes beyond Blake's celebration of childhood, encouraging the reader to open their imagination and experience the beauty, as well as the real truth – no matter how harsh it may be – that exists in the natural world.

As a final point, and in relation what constitutes Hughes's idea of childhood and how poetry should be written, Marianne Moore's much quoted definition of poetry as 'imaginary gardens with real toads in them' (1968: 266–267), seems particularly apt to describe Hughes's own poetic universe. That Hughes's work respected and depicted that realness of animals goes without saying. However, going beyond this, from an ecocritical reading, his understanding of animals depicts, using Moore's words, 'the raw material of poetry in all its rawness' (1968: 266–267). Perhaps because far from depicting Arcadia and without writing down to them, Hughes offers children an understanding of real animals. By bridging the artificial gap between the human/non-human, by revealing the beauty that lies in Earth's nature, Hughes's poetry can teach us how human beings have disturbed the natural world and the lives of animals, showing us how to look at nature from the other's point of view and educating a sense of respect towards the environment.

Notes

1 According to Paul, the original text comes from the Ted Hughes Papers and is classified as 'Unidentified notes on children's writing and teaching' at the archives at Robert W. Woodruff Library, Emory University, Atlanta, Georgia, USA.

2 Indicative of this crossover between the adult/child reader is the fact that several poems that first appeared in *Season Songs* (1976) later appeared in *Moortown Diary* (1989). Different poems from *Season Songs* (1976), *Under the North Star* (1981) and *What is the Truth?* (1984) were also included by Hughes in the two editions of *New Selected Poems* (1982 and 1995).

3 Abram coined the phrase 'the more-than-human world' as a way of referring to earthly nature and is used as the subtitle of his book *The Spell of the Sensuous: Perception and Language in a More-than-Human World* (1996). Since then the term has been largely adopted by other scholars and has become a key phrase within ecopsychology and ecocriticism.

4 First published in *The Hawk in the Rain* (1957). Collected in *New Selected Poems 1957–1994* (1995). Chapter 1 (From Pembroke to 'The Burnt Fox') analyses the inspiration behind this poem.

5 From the Greek word *oikos*, the home or dwelling place.

6 Add MS 53784, The British Library, London.

7 Hughes is of course referring here to The Natural History Museum, located in South Kensington.

8 Several versions of the book exist, all with slightly different contents. Following the 1961 edition, the America version published by Bobbs-Merrill in 1973 contained 10 poems, incorporating 'My Own True Family', whilst its second publication in England by Puffin in 1977 contained 12 poems including 'Mr Fairy Grandmother'. The page references here are from Hughes's *Collected Poems for Children* (2005).

9 The trickster is a mythological figure that can shift shape within animal or human embodiment and which has long appeared in stories of creation and recreation. According to Lewis Hyde the 'trickster is a boundary-crosser' (1998: 7), implying that he often appears in the form of a traveller and is able to cross both physical and social boundaries. Anthropologist Paul Radin defines the North American Indian trickster as both a 'creator and destroyer' (1956: xxiii). In Hughes's poetic universe, crow, as a trickster figure, wears different masks, mediating between the dualities of Hughes's creative and destructive universes or the male/female, adult/child binary. Hughes uses the trickster figure to blur these dualities and to ask questions through different narratives about moral and human values found in Western society, as in *Crow* (1970).

10 Add MS 88918/121/16, The British Library, London.

11 Both *How the Whale Became* (1963) and *The Earth Owl and Other Moon People* (1963) were dedicated to Hughes's children Frieda and Nicholas, as if to affirm his

determination and paternal instinct to nurture his two children after losing their mother.

12 *The Earth Owl and Other Moon People* (1963), is the first collection of his 'Moon Poems' which Hughes continued to develop and expand over the next 25 years: *Moon-Whales and Other Moon Poems*, 1976, *Earth-Moon*, 1976 – a limited edition illustrated by Hughes himself – and a revised edition, *Moon-Whales*, 1988. A fifth book, *Moon-Bells and Other Poems*, 1978, contains poems from the previous volumes as well as other poems that would later be published in *Under the North Star*, 1981.

13 The *Ariel* that Hughes published in 1965 was not identical to Plath's original unfinished manuscript. He was later heavily criticised for omitting poems that were to do with their marriage and which were aimed at Hughes. See Marjorie Perloff's 'The Two *Ariels*: The (Re)making of the Plath Canon' (1984), pp. 10–18.

14 Incidentally Sylvia Plath died five months before the poem 'Foxgloves' first appeared on 27 July 1963 in the *New Yorker* and in November that same year in *The Earth Owl and Other Moon People*. It was later published in *Moon-Whales and Other Moon Poems*, 1976.

15 There is another less discussed dimension to Hughes and Baskin's collaboration: the economic side to it. In a letter to Keith Sagar, dated 30 August, 1979, Hughes mentions that there 'lies a contract with Leonard to write three books of verse with his illustrations, for which he desperately needs the cash' (Sagar, 2012: 78).

16 The poems are collected here for the first time, with the exception of the 'Autumn' poems, which had previously been published as *Five Autumn Songs*, in 1968. *Season Songs* was in fact first published as a Rainbow Press edition under the title *Spring, Summer, Autumn, Winter* in 1974.

17 Baskin moved with his family to Lurley Manor, near Tiverton, Devon, where he lived from 1974 to 1983, in order to be closer to Hughes.

18 Gerald Hughes also comments on this fact: 'Ted never really gave up on his hopes that I might one day return to live in England, and particularly his idea that the ideal life for us would be running a farm together' (2012: 177).

19 I draw here on the title of Craig Robinson's book *Ted Hughes as Shepherd of Being* (1989).

20 The earliest record of the rhyme is in *Tommy Thumb's Pretty Song Book*, published in 1744. The traditional rhyme which has different versions is often used satirically in popular fiction as a murder archetype or as a parody to criticise something and bring it to public attention.

21 It was no coincidence either that Hughes's deepest discussion of the healing powers of imagination appears in his essay 'The Hanged Man and the Dragonfly', written as a tribute to Leonard Baskin in 1983, in which he describes the transformation of the horror of the hanged man into the beauty of the dragonfly. Whilst Baskin's hanged man symbolises pain and agony, the dragonfly depicts the transformation that it has undergone, and with it redemption. Baskin's transformation can be read as parallel to that of Hughes's own healing quest and search for redemption.

22 'Amulet' was first published in *Moon-Bells and Other Poems* (1978) and later reprinted as the prefatory poem in *Under the North Star* (1981).

23 The sense of 'anotherness' is used here as an alternative to the notion of 'otherness' more commonly used in ecocriticism. This notion is derived from the thought of Mikhail Bakhtin's early essay 'Toward a Philosophy of the Act' (1993) and was applied to ecocriticism by Patrick D. Murphy. Murphy goes beyond the concept of 'the other' to the notion of 'anotherness', which as he asserts is 'based on *the Another* – not the Alien and not the Stranger, but the brother, the cousin, the sister, and not just the human ones, but all the creatures with whom we share the planet' (2006: 419).

24 *Earth Dances* (1993) was the most lavish production of Lloyd's, printed in a limited edition of 250 prints by the Old Stile Press, using fine handmade papers by Frances McDowall.

25 In 1979 Hughes won the award for *Moon-Bells and Other Poems*, and in 1983 for editing with Seamus Heaney *The Rattle Bag*.

26 It is interesting to note that the poems that appeared in the collection were later republished without the prose story in Ted Hughes's *Collected Poems for Children* (2005).

27 Add MS 88918/1/59, The British Library, London.

28 The term 'poetic truth' is a concept that inhabits Graves's own interviews and essays from the 1930s. According to Graves it is 'a sort of supra-rational truth', and one that 'comes very close to the natural faith of religion' (Wheldon, 1959: 55).

29 The term comes from a letter Blake wrote to Thomas Butt, 22 November 1802 in Geoffrey Keynes (ed.), *The Letters of William Blake* (1956) 'May God us keep from single vision & Newton's sleep'.

30 In *Collected Poems for Children* (2005) it appears under the title of 'Somebody'.

31 In the summer of 1962, after having moved with Hughes to Devon in September 1961, Plath began keeping bees, as related in Plath's famous five bee poems that appear in *Ariel* ('The Bee Meeting', 'The Arrival of the Bees', 'Stings', 'The Swarm' and 'Wintering'). The poems were conceived as a sequence in less than a week in October 1962, only months before her death in February 1963. For Plath beekeeping became an analogy for the writing of poetry, and was more than a hobby. Whilst her marriage to Hughes dissolved, Plath found herself caught in a trap of post-war expectations of domesticity, which can be seen in her metaphoric bee poems.

32 Of the 28 poems, 26 reappeared in *The Iron Wolf* (1995). The other two, 'Dog' and 'Hedgehog', had already appeared in *What is the Truth?* (1984)

33 Despite the poet and artist's intentions to publish the collection they were unable to sell the books and eventually managed to convince Faber to publish them, although as Hughes writes 'Faber had about 2,000 Cats and Cuckoos for two years and never sold one. Eventually Reg and I took them back. He's selling his in fifties to art galleries, National Trust shops, etc.' (Skea, 2012: n.p.). The fact that the books never really sold perhaps endorses the fact that the rhymes are not Hughes's best.

34 MSS 644, Box 115, FF20. Robert W. Woodruff Library, Emory University, Atlanta, Georgia, USA.

35 Hans Christian Anderson's fairy tale tells of how a young mermaid is willing to give up her life at sea and her identity in order to gain a human soul and the love of a human prince. However, it is also a tale of redemption, since when the prince marries the Little Mermaid's heart breaks. When the Sea Witch offers to turn the Little Mermaid back into a mermaid again if she kills the prince, she is unable and throws herself into the ocean and her body dissolves. According to Bradley Shewaga, the fairy tale mirrors Plath's own life and her death. Shewaga's fascinating, although somewhat far-fetched essay claims that, 'The story may have represented a subconscious model for Plath, a model influencing the poor decisions that she made in adulthood, ultimately leading to her demise' (2011: 106). Shewaga parallels Plath's own life to that of the mermaid's idealisation with the prince, reminding us that 'the prince is not the perfect unblemished statue, nor was Hughes' (2011: 109) and links Hughes's affair with Assia Wevill and his infidelities to Plath's own sense of abandonment which led to her ultimate fate.

36 Add MS 88978/1, The British Library, London.

37 Not only the diminishing number of marine species and the disruption of marine ecosystems as a result of climate change, but other ecological problems derived such as over-fishing, marine pollution and rising water temperatures, all of which should be of major concern to humans given that they have extraordinary irreversible effects on the ecosystem of our marine life, our weather, climate and ultimately ourselves.

38 John Keats, Letter to Benjamin Bailey, 22 November 1817 from *John Keats Letters* (1817–1818). Available online at: www.john-keats.com/briefe/221117.htm.

Archive material

Add MS 53784. Ted Hughes and Sylvia Plath: Verse and Prose (The British Library, Western Manuscripts, London, 1952–1962).

Add MS 88918/121/16. Tree Planting Project (The British Library, Western Manuscripts, London, 25 January 1972–18 July 1991).

Add MS 88918/1/59. What is the Truth? A Farmyard Fable for the Young (The British Library, Western Manuscripts, London, 21 July 1983–21 March 1984).

Add MS 88978/1. Ann Skea Notebooks (The British Library, Western Manuscripts, London, 9 October 1992–January 1999).

MSS 644, Box 115, FF20. Unidentified, Series 2: Writings by Ted Hughes, circa 1949–1990, Subseries 2.3d: Essays, Reviews, Talks from Ted Hughes Papers, 1940–1999 (Robert W. Woodruff Library, Emory University, Atlanta, Georgia, USA).

References

Abram, David, *The Spell of the Sensuous: Perception and Language in a More-than-Human World* (New York: Vintage Books, 1996).

Adams, John, 'Dark Rainbow: Reflections on Ted Hughes', in Nancy Chambers (ed.), *The Signal Approach to Children's Books* (Harmondsworth: Kestrel Books, 1980), pp. 101–108.

Aftandilian, Dave (ed.), *What Are the Animals to Us?: Approaches from Science, Religion, Folklore, Literature, and Art* (Knoxville: University of Tennessee Press, 2007).

Auden, Wystan Hugh, *A Choice of de la Mare's Verse* (London: Faber & Faber, 1963).

Bakhtin, Mikhail, *Toward a Philosophy of the Act* (Texas: University of Texas Press, 1993).

Bate, Jonathan, *The Song of the Earth* (London: Picador, 2000).

Berger, John, *About Looking* (New York: Vintage International, 1980).

Bryson, J. Scott, *Ecopoetry: A Critical Introduction* (Salt Lake City: University of Utah Press, 2002).

Buell, Lawrence, *Writing for an Endangered World: Literature, Culture, and the Environment in the U.S. and Beyond* (Cambridge: Harvard University Press, 2001).

Carson, Rachel, *Under the Sea-Wind* (New York: Oxford University Press, 1941; centennial edition, New York: Penguin, 2007).

Carson, Rachel, *The Sea Around Us* (New York: Oxford University Press, 1951; new edition, Oxford: Oxford University Press, 1991).

Carson, Rachel, *Silent Spring* (Boston, MA: Houghton Mifflin, 1962; London: Penguin, 1999).

Carson, Rachel, *The Sense of Wonder* (New York: Harper & Row, 1965; New York: HarperCollins, 1998).

Chanan, Noel (dir.), *The Artist and the Poet: Leonard Baskin & Ted Hughes in Conversation 1983*. DVD. 2009.

Clark, Timothy, *The Cambridge Introduction to Literature and the Environment* (Cambridge: Cambridge University Press, 2011).

Cook, Peter and Mick Gowar, 'Within the Hearing, and Seeing, of Children', *The Ted Hughes Society Journal*, 1 (Summer 2011). Web. 20 July 2013. www.thetedhughessociety.org.

Corrin, Stephen, Letter published in *Signal: Approaches to Children's Books*, 30 (September 1979), p. 169.

Cushman, Keith, 'Hughes' Poetry for Children', in Keith Sagar (ed.), *The Achievement of Ted Hughes* (Manchester: Manchester University Press, 1983), pp. 239–256.

Dinsdale, Tim, *Loch Ness Monster* (London: Routledge & Kegan Paul, 1961).

Elder, John, *Imagining the Earth: Poetry and the Vision of Nature* (Athens: University of Georgia Press, 1996).

Eliade, Mircea, *Shamanism: Archaic Techniques of Ecstasy* (Bollingen Series, 76. New York: Pantheon Books, 1964).

Gifford, Terry (ed.), *The Cambridge Companion to Ted Hughes* (Cambridge: Cambridge University Press, 2011).

Gifford, Terry and Neil Roberts, *Ted Hughes: A Critical Study* (London: Faber & Faber, 1981).

Gowar, Mick, '*Moon-Bells and Other Poems*', *Ted Hughes Society*, 2011. Web. 10 July 2015. www.thetedhughessociety.org/moonbells.htm.

Graves, Robert, *The White Goddess: A Historical Grammar of Poetic Myth* (London and Boston: Faber & Faber, 1948); New York: Farrar, Straus & Giroux, 1966).

Ham, Jennifer and Matthew Senior (eds), *Animal Acts: Configuring the Human in Western History* (London: Routledge, 1994).

Hughes, Gerald, *Ted and I* (London: The Robson Press, 2012).

Hughes, Ted, *The Hawk in the Rain* (London: Faber & Faber, 1957).

Hughes, Ted, '"Five Ton Phantom", Review of *Loch Ness Monster* by Tim Dinsdale', *New Statesman* (2 June 1961), p. 61.

Hughes, Ted, *Meet My Folks* (London: Faber & Faber, 1961).

Hughes, Ted, *How the Whale Became* (London: Faber & Faber, 1963).

Hughes, Ted, *The Earth-Owl and Other Moon-People* (London: Faber & Faber, 1963).

Hughes, Ted, *Nessie the Mannerless Monster* (London: Faber & Faber, 1964).

Hughes, Ted, *Poetry in the Making* (London: Faber & Faber, 1967).

Hughes, Ted, *Wodwo* (London: Faber & Faber, 1967).

Hughes, Ted, *Crow: From the Life and Songs of the Crow* (London: Faber & Faber, 1970).

Hughes, Ted, *Season Songs* (London: Faber & Faber, 1976).

Hughes, Ted, *Moon-Whales and Other Moon Poems* (New York: Viking, 1976).

Hughes, Ted, *Earth-Moon* (London: Rainbow Press, 1976).

Hughes, Ted, *Cave Birds* (London: Faber & Faber, 1978).

Hughes, Ted, 'Norwich Tapes Ltd, London: *The Critical Forum*', 1978. Web. 31 August 2013. http://ann.skea.com/CriticalForum.htm.

Hughes, Ted, *Moon-Bells and Other Poems*, (London: Faber & Faber, 1978).

Hughes, Ted, *Moortown* (London: Faber & Faber, 1979).

Hughes, Ted, *Under the North Star* (London: Faber & Faber, 1981).

Hughes, Ted, *What is the Truth?* (London: Faber & Faber, 1984).

Hughes, Ted, *The Cat and the Cuckoo* (London: Sunstone Press, 1987).

Hughes, Ted, *Moon Whales* (London: Faber & Faber, 1988).

Hughes, Ted, *Moortown Diary* (London: Faber & Faber, 1989).

Hughes, Ted, *The Mermaid's Purse*, Illus. R. J. Lloyd (Bideford: Sunstroke Press, 1993).

Hughes, Ted, *Winter Pollen* (London: Faber & Faber, 1994).

Hughes, Ted, *Earth Dances*, Illus. Reg Lloyd and R. Monmouth (Gwent: The Old Stile Press. 1994).

Hughes, Ted, *New Selected Poems 1957–1994* (London: Faber & Faber, 1995).

Hughes, Ted, 'Preface', in *R.J. Lloyd: A Retrospective Exhibition of Selected Works* (Hereford: Hereford City Museum and Art Gallery, 1996), pp. vi–vii.

Hughes, Ted, *Birthday Letters* (London: Faber & Faber, 1998).

Hughes, Ted, *Collected Poems for Children* (London: Faber & Faber, 2005).
Hughes, Ted and Seamus Heaney (eds.), *The Rattle Bag* (London: Faber & Faber, 1982).
Hyde, Lewis, *Trickster Makes This World: Mischief, Myth, and Art* (New York: Farrar, Straus & Giroux, 1998).
Kalof, Linda, *Looking at Animals in Human History* (London: Reaktion Books, 2007).
Kazzer, Claas, 'Earth Moon: Ted Hughes's Books for Children (& Adults)', in Joanny Moulin (ed.), *Ted Hughes: Alternative Horizons* (London: Taylor & Francis, 2004), pp. 111–124.
Keats, John, 'Letter to Benjamin Bailey. 22 November 1817 from *John Keats Letters (1817–1818)*'. Web. 20 September 2014. www.john-keats.com/briefe/221117.htm.
Keynes, Geoffrey (ed.), *The Letters of William Blake*, 1956. Web. 5 May 2014. https://archive.org/details/lettersofwilliam002199mbp.
Kowalski, Gary A., *The Souls of Animals* (Walpole, NH: Stillpoint, 1991).
Lockwood, Michael, 'Ted Hughes: The Development of a Children's Poet', *Children's Literature in Education*, 40:4 (December 2009), pp. 296–305.
Malamud, Randy, *Poetic Animals and Animal Souls* (New York: Palgrave Macmillan, 2003).
Merchant, Carolyn, *Radical Ecology* (New York: Routledge, 1992).
Midgley, Mary, *Animals and Why They Matter* (Athens: University of Georgia Press, 1984).
Moore, Marianne, 'Poetry', in *The Complete Poems of Marianne Moore* (London: Faber & Faber, 1968), pp. 266–267.
Moses, Kate, 'Sylvia Plath's Voice Annotated', in Anita Helle (ed.), *The Unraveling Archive: Essays on Sylvia Plath* (Ann Arbor: University of Michigan, 2007), pp. 89–117.
Motion, Andrew, 'Foreword', in Morag Styles, Louise Joy and David Whitley (eds), *Poetry and Childhood* (Stoke-on-Trent: Trentham Books, 2010), p. ix.
Murphy, Patrick D., 'Grounding Anotherness and Answerability through Allonational Ecoliterature Formations', in Catrin Gersdorf and Sylvia Mayer (eds), *Nature in Literary and Cultural Studies: Transatlantic Conversations on Ecocriticism* (New York: Rodopi, 2006), pp. 417–434.
Neill, Heather, 'The Nature of Poetry', *Times Educational Supplement Magazine*, 2 June 1995. Web. 20 June 2013. www.tes.co.uk/article.aspx?storycode=10719.
Paul, Lissa, '"Writing Poetry for Children is a Curious Occupation": Ted Hughes and Sylvia Plath', *The Horn Book Magazine*, 81:3 (May/June 2005), pp. 257–267.
Perkins, Agnes, 'A Striking Contrast: Recent British & American Poetry for Children', *Children's Literature*, 10:1 (1982), pp. 186–193.
Perloff, Marjorie, 'The Two *Ariels*: The (Re)making of the Plath Canon', *American Poetry Review*, 13:6 (November/December 1984), pp. 10–18.
Philip, Neil, 'The Shawl of the Beauty of the World: The Children's Books of Ted Hughes', *Signal: Approaches to Children's Books*, 100 (2003), pp. 191–202.
Plath, Sylvia, *Ariel* (London: Faber & Faber, 1965) Plath, Sylvia, *Ariel* (London: Faber & Faber, 1965).
Plath, Sylvia, *Letters Home: Correspondence 1950–1963*, edited by Aurelia Plath (London: Faber & Faber, 1976).
Potter, Beatrix, *The Tale of Jemima Puddle-Duck* (London: Frederick Warne, 1908; London: Penguin, 1989).
Pratt, C. (ed.), *An Encyclopedia of Shamanism* (New York: The Rosen Publishing Group, 2007).
Radin, Paul, *The Trickster* (London: Routledge, 1956).

Reid, Christopher (ed.), *Letters of Ted Hughes* (London: Faber & Faber, 2007).
Roberts, Neil, *Ted Hughes: A Literary Life* (Basingstoke: Palgrave Macmillan, 2006).
Roberson, Ed, 'We Must Be Careful', in Camille Dungy (ed.), *Black Nature* (Athens: University of Georgia Press, 2009), pp. 3–5.
Robinson, Craig, *Ted Hughes as Shepherd of Being* (Basingstoke: Macmillan, 1989).
Sagar, Keith, (ed.), *The Art of Ted Hughes* (Cambridge: Cambridge University Press, 1978).
Sagar, Keith (ed.), *The Achievement of Ted Hughes* (Manchester: Manchester University Press, 1983).
Sagar, Keith (ed.), *Poet and Critic: The Letters of Ted Hughes and Keith Sagar* (London: British Library, 2012).
Shewaga, Bradley, 'Sylvia Plath: The Woman Who Gave Up Her Voice', *Plath Profiles: An Interdisciplinary Journal for Sylvia Plath Studies*, 4 (June 2011), pp. 106–114.
Sideris, Lisa and Kathleen Moore (eds), *Rachel Carson: Legacy and Challenge* (New York: SUNY, 2008).
Skea, Ann, 'Ted Hughes and the Goddess', *The Ted Hughes Society*, 2012. Web. 22 January 2013. www.thetedhughessociety.org/thegoddess.htm.
Skea, Ann, 'A Creative Collaboration: R. J. Lloyd and Ted Hughes', 2012. Web. 24 August 2013. http://ann.skea.com/ArtisticCollaboration1.htm.
Skinner, Jonathan, 'Editor's Statement', *ecopoetics*, 1 (Winter 2001), pp. 5–8. Web. 8 May 2014. www.factoryschool.org/ecopoetics/.
Starhawk, *The Spiral Dance* (San Francisco: Harper, 1979; 2nd edition, 1989).
Stokes, John, '"Children are Obsessives", Review of *Season Songs* by Ted Hughes', 1976. Web. 30 March 2012. http://english.oxfordjournals.org/.
Styles, Morag, *From the Garden to the Street: Three Hundred Years of Poetry for Children* (London: Cassell, 1998).
Styles, Morag, 'Poetry', in Peter Hunt (ed.), *International Companion Encyclopedia of Children's Literature* (London: Routledge, 2004), pp. 396–417.
The English Programme, Passwords: Ted Hughes, Double Exposure North for Channel 4, Passwords, Programme 3, The English Programme (Channel 4 Schools video, 1998).
Townsend, John Rowe, *Written for Children: An Outline of English Language Children's Literature* (Harmondsworth: Kestrel Books, 1983).
Underhill, Hugh, *The Problem of Consciousness in Modern Poetry* (Cambridge: Cambridge University Press, 1992).
Wain, John, Citation to 'The Winner of the Signal Poetry Award', *Signal: Approaches to Children's Books*, 29 (May 1979), pp. 63–66.
Walder, Dennis, *Ted Hughes* (Milton Keynes: Open University Press, 1987).
West, Thomas, *Ted Hughes* (London: Methuen, 1985).
Wheldon, Huw P., 'Robert Graves', in Frank L. Kersnowski (ed.), *Conversations with Robert Graves* (Jackson: University Press of Mississippi, 1959), pp. 49–56.
Whitley, David, '"The Fox is a Jolly Farmer and We Farm the Same Land": Ted Hughes and Farming', in Mark Wormald, Neil Roberts and Terry Gifford (eds), *From Cambridge to Collected* (Basingstoke: Palgrave Macmillan, 2013), pp. 96–111.
Wilson, Anthony, 'Ted Hughes's Poetry for Children', *Children's Literature in Education*, 32:2 (2001), pp. 77–90.
Wilson, Edward O., *Biophilia* (Cambridge, MA and London: Harvard University Press, 1984).

6 Hughes's children's prose

Hughes's literary career took place parallel to the growth of his environmental awareness: from the poet's realisation about the polluted state of the rivers of his childhood (The Calder and the Don), to his reading of Rachel Carson in July 1958, followed by his farming practices and concerns about pollution, climate change and species extinction during the 1970s and 1980s. As a result, different social and political interventions related to environmental concerns inform his creative work underlying his sense of social ecology and ecological responsibility.[1]

Given that Hughes's approach to ecology is largely a holistic one his environmental responsibility and 'ecological sensibility' celebrate imagination as an important factor to change human's way of thinking. Hughes's goddess of nature, ecofeminism, deep ecology and social ecology all call attention to understanding the relationship between nature and culture as a holistic unity. Such an approach to the nature/culture split would appear to be very much consistent with Hughes's own healing quest and his poetic responsibility to reconnect human beings with nature. It is precisely this social interaction with nature together with Hughes's concern and engagement with environmental issues that are also social issues that underlie Hughes's healing myths *The Iron Man* (1968) and *The Iron Woman* (1993), both of which are largely informed by Hughes's own social and political interventions related to environmental concerns such as river pollution and the dumping of toxic waste.

The healing function of literature, the idea that it can makes us emotionally and physically stronger, goes back to Aristotle and Plato. It was Aristotle in his *Poetics* who first used the term *catharsis* referring to the mental purification or purging of the emotions. This holistic sense of literature – the idea that storytelling can make us more 'resilient' and help us heal – is very much part of Hughes's own mythical storytelling and key to understanding his healing quest. As both a myth-maker and a storyteller, Hughes created stories that capture the essence of experience through the affective function of language. This chapter focuses on Hughes's most representative works in prose with the exception of his picture books for young children, *Shaggy and Spotty* (1997), illustrated by David Lucas, and *Timmy the Tug* (2009), by Jim Downer and Ted Hughes.[2] For it is through the art of storytelling, and especially in *The Iron Man* (1968) and *The Iron Woman* (1993), where Hughes's healing quest is most clearly achieved.

The redemptive voice of pain: *Ffangs the Vampire Bat and the Kiss of Truth (1986)*

Ffangs the Vampire Bat and the Kiss of Truth (1986) was written just two years after Hughes had been appointed Poet Laureate and published *What is the Truth?* (1984). Like *What is the Truth?* the book takes on the form of an animal fantasy, with a moralistic tone, but is more light-hearted. The plot begins by narrating how Attila, a cockerel, is transformed from a rooster to a vampire bat. Paradoxically whilst *What is the Truth?* moves from prose to verse, *Ffangs* begins in rough verse and later drifts into prose. Technically it resembles *Nessie the Mannerless Monster* (1964), and follows a similar thematic development. Despite his name, Ffangs is not a real vampire, but a 'failed vampire' (*FTV*: 34), who can't stand the sight of blood. Like Nessie he is an outsider excluded from society and roams the streets wishing 'to be human and happy' (*FTV*: 39). In a similar way, Ffangs's quest to find 'the truth' requires that the monster must not only be confronted but integrated into society. The story continues with Squarg, a vampire hunter, who is called in from Transylvania, and plans to attract Ffangs by using Sweety Crisp, a beauty queen. However, Squarg's plan backfires, and whilst Squarg marries Sweety Crisp, Ffangs becomes their best man, Ffangs is then set up in a restaurant, where, in Hughes's absurd humour, he 'feeds his customers with oozing steaks and livers and stuffed hearts' whilst he 'buries his furry muzzle in strawberries and cream' (*FTV*: 71). Hughes's peculiar story could, and perhaps should, have finished there, but instead we are led in the following chapter into the tale of Squarg trying to find the truth about his wife Sweety Crisp who has been turned into a rat. The story then takes yet another twist when we learn that the rats had cast a spell on his wife who was a beautiful vampire and had been changed into the woman he married. The only person who could undo the spell was one 'who could make time go backwards' (*FTV*: 102). Help eventually comes from an old woman who turns a silver bullet into a snake and tells them to follow it. The last chapter draws on the subtitle of the book 'The Kiss of Truth', where Ffangs is transformed into a boy, Bright Lad, after being kissed by the 'truth' in the form of the green snake, which has the power to regenerate love. And so it is at the end of Ffang's quest, after confronting the monster, that he is finally restored to human form. The abrupt ending concludes with a note that appears on the last page announcing that 'The adventures of Ffangs will be continued in the next book' (*FTV*: 96).[3] However, most certainly due to its poor reception, Hughes never wrote a sequel. This abrupt ending becomes even more relevant bearing in mind the strong autobiographical aspect underlying the book. *Ffangs* was, for Hughes, something of an emotional and psychological test to surpass, an attempt to produce a personal redemptive story, one which he would only achieve in his play 'Orpheus' or later in his adult work in *Birthday Letters* (1998).

Another reason why the book renders as being somewhat confusing to the reader is largely due to the lack of flow in transition from place to place – one minute Ffangs is in London at the Houses of Parliament, or at Buckingham

Palace talking to the Queen, whilst the next he is on the moon. Hughes also makes a rather liberal use of vampire mythology with Ffangs desire to become human throughout.

According to Hughes, unlike other books, writing *Ffangs* was not a premeditated decision but something that came by accident, as he explains in a letter to Nick Gammage in 1989,

> I wanted to demonstrate (to a close friend) what seem to me the basic simple laws of a narrative – and so wrote the first chapter … the book was designed as a sort of gymnasium – for reading ability in the very young.
>
> (*LTH*: 572)

Despite Hughes's confession, bearing in mind that, as a writer, Hughes was not one to do things 'by accident', it would appear that the story represents a somewhat unsuccessful attempt to write about his most intimate self as part of his healing narrative. As Claas Kazzer correctly points out it is here that 'Hughes writes most openly about the kind of Truth most closely associated with the goddess, and closest to his intentions in writing. It is the Truth of finding out who one really is' (2004: 123). More recently Jonathan Bate has also made the claim that, before writing *Birthday Letters*, in *Ffangs* Hughes was 'approaching his own story indirectly' (2015: 430) and that 'There was more of that true self in this story than in the poetry collections he published at this time' (2015: 430). According to Bate, Selena, the protagonist, 'is a figuration of the White Goddess, but her bite-kiss which transforms Ffangs into his true self is also that of Sylvia at Falcon Yard' (2015: 430).[4]

Was Hughes attempting in *Ffangs* to create a fictional double of himself and Plath in an attempt to tell his own story indirectly? Whilst Bate's idea is tempting it does seem a rather far-fetched speculation. Selena is of course a figuration not only of Plath as Hughes's muse, but also of Graves's *White Goddess* (1948), which like Plath, haunted Hughes throughout his entire life. Hughes's goddess, like that of Robert Graves, is both desirable and fearful, for she 'will suddenly transform herself into sow, mare, bitch, vixen, she-ass, weasel, serpent, owl, she-wolf, tigress, mermaid or loathsome hag. Her names and titles are innumerable…' (Graves, 1966: 24).

In Hughes's works the goddess, in her different manifestations, appears as both a benign and dark element. Whilst she appears as the 'bride' in *Crow* (1970), in *Ffangs* she takes on first the form of a hag, who then turns into a beautiful maiden whom Ffangs rescues. In both stories the female is associated with the snake. In *Ffangs* she is depicted in the character of Selena, who is found on the moon with a snake in her mouth, and whose bite changes people into what they really are. That Hughes chooses the name Selena is not incidental either, given that in Greek mythology Selene was the mother goddess, linked to the full moon, pale faced and with black hair, who in this case changes her moon chariot for a ride on the back of the Vampire.[5] As Hughes explains in the tale 'When Thomas asked the girl her name, she told him it was Selena.

"It means Moon", she shouted, against the rushing slipstream of the Bat's flight. "That's why I love sunbathing. And that's why my hair's jet black"' (*FTV*: 139).

Whether or not Hughes was attempting to story his and Plath's first encounter, judging from the unfavourable reception, it appears to have been a failed attempt. *Ffangs* received very little mention by critics and has remained largely ignored, perhaps due to the lack of intimacy in the main characters and the constant change of perspective which makes the plot appear disorganised and hard to engage with. Whilst critics like Neil Roberts have claimed that *Ffangs* is 'perhaps the strangest of Hughes's books for children ... unambiguously addressed to children, it is a chaotic riot of narrative' (2006: 176), other scholars, such as Andrew Armitage (2010) have suggested that the symbol of the bat corresponds with Plath's description of Hughes as 'the vampire' in 'Daddy' who 'drank [her] blood for seven years' (Plath 1981: 224), and is likely to have prompted his children's book.[6] Despite the somewhat fractured and rambling plot, the book's real interest can be found in reading it as Hughes's confrontation with the darkness of the human psyche and his need for self-expression, only available to the poet through the realm of his children's writing. As he explained to Blake Morrison 'When you're writing for adults you feel you have to smuggle it past a tremendously vigilant defence system' (1999: 160).[7] For Hughes storytelling was the means to make sense of the world, and storying his own life was part of his quest as a writer to explore the boundaries between reality and fantasy. Writing for children meant that Hughes was able to let his 'defence system' down, and touch on his own private life. As he confesses:

> I remember a little sequence I wrote for children, *Ffangs the Vampire Bat*, which was sent for review to a poet I know, who wrote to me and said, 'How can you expose yourself like that?' He'd put his finger right on it. That's why you write: to feel your secret life leaking out in places.
>
> (Morrison, 1999: 160)

It makes sense that Hughes had had high hopes for *Ffangs* and was devastated when it was not received as well as he had expected. Its unfortunate reception would also explain why he never finished it, despite the fact that there are proofs of a different version to *Ffangs*, dating from 1979, and at least one unpublished draft of '*Ffangs 2*'.[8] However, perhaps the real problem with *Ffangs*, and the reason why it fails to deliver, is the fact that it denies its readers something which Hughes himself considered a key component in any children's story: a true 'happy ending'. This would appear to be due to the fact that in *Ffangs* the personal aspect tends to dominate the story and so, from an ecocritical reading, the human relation with nature is neglected. Notwithstanding, if writing entailed, as Hughes himself considered, 'trying to take fuller possession of the reality of your life' (Morrison, 1999: 171), then *Ffangs* can be considered as one of Hughes's most intimate and personal narratives, since in creating *Ffangs*, the poet was creating his own character by storying his past and enacting his own role.

To understand the importance of Hughes using his own self as a character within the story there is a curious incident worth mentioning here briefly. It is important to remember that Hughes was deeply superstitious and so when he found a bat on Boston common, whilst living in America with Sylvia Plath, he considered it as an omen. In a letter to Baskin, in July 1959, Hughes tells how he found the bat under a tree: 'I put it back on the tree trunk & it went scrambling away upwards, upside down, snarling downwards and shrieking. Its teeth just pricked my finger in one place, so now I'm waiting for rabies' (*LTH*: 147). The occasion must have been significant enough for Hughes to recall in the poem '9 Willow Street' where he narrates the event, entwining fiction with facts, revealing Hughes's own emotional state of pain and the guilt he felt after Plath had killed herself.

That Hughes's persona in *Ffangs* chose to be a bat may have been coincidence, although in somebody who took astrology and the occult powers so seriously, it is doubtful that he would have not made connections with the 'bat episode' from his past. What is perhaps more interesting is the fact that he chose to expose himself in such a way in his children's writing rather than in writing for adults where he felt more vulnerable. It is precisely that sense of vulnerability of our inner child-self as part of the human condition that Hughes speaks of in 1986, in a very personal letter to his son Nicholas, 24, at the time:

> It's something people don't discuss, because it's something most people are aware of only as a general crisis of sense of inadequacy, or helpless dependence, or pointless loneliness ... But not many people realise that it is, in fact, the suffering of the child inside them.
>
> (*LTH*: 513)[9]

Read in the context of the poet's own personal tragedy, Hughes's confessional letter to Nicholas illuminates the spiritual importance of keeping in touch with the child within. In it the poet talks about not only how he understands the child's sensibility but how it informs his work. Hughes appears to be referring here, once again, to Carson's 'sense of wonder' and the crossover barrier between the child-adult reader, the fact that the poet has never lost touch with his inner child-self. Throughout his life Hughes was desperate to protect not only his own privacy, but also to shield his children from the public eye, to the point that he became, not in the words of his fiercest critics, but in his own, a 'monster of self-defence' (*LTH*: 246). However, there was also the need to let down the barriers and turn off those mechanisms of self-defence in order to embrace the child within. It is in works like *Ffangs*, that the poet attempts to do just this.

In this light *Ffangs* can be read as a kind of reconciliatory story, a redemptive story of self-acceptance and of coming to terms with one's real self. Hughes's storytelling and his own storying was a way of finding a world where he could write with honesty and openness. Seen from this perspective, was Hughes writing *Ffangs* in the context of representing the state of his relationship with

Sylvia Plath? Was it by confronting his own demons, and by looking back in retrospect, that he was able to make sense of his own past? Hughes's tortured battles with a split creative/destructive goddess, together with the personal tragedies of the deaths of Sylvia Plath and Assia Wevill would appear to justify calling *Ffangs* a story of redemption at least, and one that belongs to the inherent healing of Hughes's poetic myth-making.

Hughes's creation tales

How the Whale Became (1963), Hughes's first collection of stories, was written and published by Faber and Faber in 1963 and illustrated by George Adamson. Faber republished the book in 2000, and it was later reissued in paperback by Frances Lincoln in 2008, accompanied by award-winning artist Jackie Morris's evocative illustrations that so gracefully capture the imagery and spirituality of Hughes's animal tales.

Although Hughes began writing these tales around the age of 26, he continued to draft and write his 'creation tales' about how the world and its creature came into being, throughout his life. *Tales of the Early World* was published in June 1988, followed by *The Dreamfighter and Other Creation Tales* in March 1995.[10] However, whilst preparing materials to be put on sale in the mid-1990s, Hughes kept back a box folder which contained notebooks of drafts of more creation tales to work on, as well as a list of over 50 stories, which he had intended to write, to the point that these 'myths of origin were a peculiar obsession throughout his writing career' (Bate, 2015: 120).[11] Indeed Hughes found the fable a congenial form to use at different stages of his writing career. Throughout his creation tales, Hughes appears to use the same pattern and mental framework to build the different stories which, in the annotated versions that can be found in the archives, first appear as mind maps on a blank page in preparation to creating notes which he later used to draft the different moral fables. Despite the fact that originally fables had been intended for an adult public and often had hidden political statements, until the mid-eighteenth century they were the only approved secular reading for the young and used to teach ethics. Given that, as a genre, fables belong to the tradition of children's literature, and have been traditionally classed as didactic stories with a moral purpose, it suited Hughes to use this format for his tales, in order to provide stories with both a problem and a solution. According to the writer William Golding 'The fabulist is a moralist. He cannot make a story without a human lesson tucked away in it' (1977: 227). Hughes's ideas appear to be very much in line with Golding,[12] who believed that Western culture had created unhealthy, protective taboos and that the fabulist was didactic and desired to introduce a moral lesson into an imaginary story. However, as Golding cautions, this puts the fabulist into an awkward position since:

> People do not much like moral lessons. The pill has to be sugared, has to be witty or entertaining, or engaging in some way or another … if the pill is

> not sufficiently sugared it will not be swallowed. If the moral is terrible enough he will be regarded as inhuman; and if the edge of his parable cuts deeply enough, he will be crucified.
>
> (1977: 228)

This is largely related to the fact that children do not read fables in the same way as adults. Whilst adults may find certain fables disturbing and unsuitable for children, a child is more likely to understand that the story does not impose any potential threat. In writing fables for both children and adults Hughes exposes our hidden self, provoking the reader, without patronising and sheltering the child. An example of this can be seen in how the repeated themes of pride, envy, greed, violence and cruelty that appear in his tales are used to reinforce survival rather than death.

How the Whale Became *(1963)*

In the summer of 1956 Plath and Hughes visited Paris, and later Spain, together, as a newlywed couple. They arrived in Madrid with little more than a rucksack and Sylvia's typewriter and went on to Benidorm, which at the time was a small fishing village where, during the hot months of July and August, Hughes, at the age of 26, wrote his first animal fable 'How the Donkey Became' which was later to become part of *How the Whale Became*. The collection consists of 11 prose fables, explaining the origins of different animals and how the creatures acquired their present forms. According to Hughes, they 'were all written with a moral purpose, in Aesop style' (1992: 147).

In a letter Hughes wrote to his sister Olwyn in August, 1956, he claims that 'It's nearest model is Swift, I'm sure. But much simpler' (*LTH*: 46). Hughes's admiration for Swift's style stemmed mainly from the 'clarity, precision, conciseness and power' (*LTH*: 20) that Hughes saw in his work. Hughes's tales share a similar ironic dark sense of humour, closer to that of Swift than the light-hearted comic style of Kipling's *Just So Stories*, which would have seemed a more obvious model of inspiration, given the fact that both are written as fables, and follow the conventional style of myths of origin (*pourquoi* tales).[13] Kipling's tales are clearly modelled on the story frame of the *Jatakas*,[14] a collection of ancient Indian fables, which Hughes would also have been familiar with. Despite Hughes's claim that his tales are moralising, the fables read as being more subversive than Kipling's folkloric style, which often reads as poetic prose and have a more perceptive sense of wit. Hughes often provokes the young reader indirectly to make sense of what is right and wrong by using animal characters to reflect human behaviour in their quest to establish identity.

Like Kipling's *pourquoi* tales Hughes uses animals to portray human vices, such as selfishness, laziness or vanity, amongst others. However, whilst Kipling's tales have a precise underlying geographical reference (Africa, the Far East, America and ancient Europe) and were written with the didactic intention of educating the reader into the politics behind the Empire, Hughes's tales lack any particular geographical reference. As Neil Roberts points out, they share few other similarities 'Perhaps the most interesting difference, however, is that God plays no explicit part in Kipling's creation, but is a central figure in Hughes' (2006: 171). Indeed, Hughes's God was to become a central character throughout his creation tales, and is depicted as being a very human kind of God, one who is prone to human weaknesses such as tiredness, lack of concentration or being absent-minded, and who alongside his successes often gets things wrong and makes a few mistakes along the way. Like Kipling's tales, Hughes takes us back to the early days of the world, when animals had not yet taken their true form and narrates how God creates the different creatures. The 11 stories tell the different ways in which his creatures come to be: Owl, Whale, Fox, Polar Bear, Hyena, Tortoise, Bee, Cat, Donkey, Hare and Elephant.

Despite being written before *Meet My Folks* (1961), *The Earth Owl and Other Moon People* (1963) and his picture book *Nessie the Mannerless Monster* (1964), all of which were published in the early 1960s, the book was not published until 1963, shortly after the birth of Frieda and Nicholas, to whom the book is dedicated. The delay in its publishing was largely due to the fact that the tales had been turned down by the British Broadcasting Corporation (BBC) Children's Department, on the grounds that they were 'too abstract and nebulous in conception and execution' (*LTH*: 81). Plath had even sent a draft of *How the Whale Became* to the Atlantic Monthly Press in Boston, through a close friend, but the tales were also rejected as not being suitable, and according to the editor 'too bloody … There is too much viciousness – the animals are portrayed as beasts rather than animals to respect and be comfortably afraid of' (Middlebrook, 2004: 98). Annoyed, but not put off, Hughes himself *did* think that they would sell, as he wrote to Gerald, a year later, claiming that the stories came to him 'absolutely naturally' and that he could produce one every day before breakfast,

> I would like to produce a classic volume – about 5,000 – children's stories. I shall bring in all the situations & characters etc out of all the fairy tales, animal tales etc that I have read & I have read millions in the last six years.
> (*LTH*: 108)

Curiously enough the most positive review came from the unlikely hand of the fashion designer Gloria Vanderbilt, who claimed that

> These 11 enchanting stories deserve to take their place among the classic fables. They come alive because they are rooted in the fundamental truth of the need for identity, and they illustrate the happiness that comes when we stop pretending to be something we are not and start being ourselves.
> (1964: 28)

The importance of identity that Vanderbilt refers to here, as a theme which is repeated throughout the tales, can be read in line with Hughes's own quest for truth. Furthermore, from a Jungian reading it would appear to correspond with the individuation process, a term created by Jung to describe the process of becoming aware of oneself and discovering one's true, inner self.

Hughes's creative power is of course best expressed through animal forms, so the fable would seem a particularly well suited form. Perhaps this is because, for Hughes

> the animals are representations of the goddess. Whatever moral implications my poetry may have can be inferred from this. The scheme behind my poetry, as I see it, and insofar as there is one, could be described as the negotiations with the goddess, outside any particular religion.[15]

The relationship Hughes depicts between nature and the animals underlines his sense of responsibility and engagement and gives insight into the laws and powers that govern Hughes's own creative and imaginative universe.

Hughes's attempt to tell the story of the creation of the world inevitably draws on Darwin's post-evolution theory, portraying evolution through the myth of nurturing Mother Earth. This can be seen in the characterisation of many of the creatures which share similarities that show signs of their common ancestry or similar traits representing the progression of species change, as in the case of the dog and the fox. Hughes's creation tales stress that everything is interconnected and review the relationship of humans and the environment, as can be seen in the opening pages, when Hughes begins the tales with his own particular version of *Genesis*:

> Long ago when the world was brand new, before animals or birds, the sun rose into the sky and brought the first day.
> The flowers jumped up and stared around astonished. Then from every side, from under leaves and from behind rocks, creatures began to appear.
> In those days the colours were much better than they are now, much brighter. And the air sparkled because it had never been used.
>
> (*HTWB*: 10)

In 'The Interpretation of Parables' (1992),[16] an article Hughes wrote in response to the claim that one of the stories in *How the Whale Became* was racist,[17] he explains how his 'fables' were written with an underlying moral purpose and how the young reader is 'led to practise and strengthen, in imaginative play, a process that can later become real, in daily life' (148). In the article Hughes responds ironically to the claim that 'To read this story as in any way "racist", because of the Polar Bear's whiteness, is to condemn it with the blindness that it is meant to immunise the reader against' (1992: 149). Each of the stories has a simple moral and is intended to 'sensitize the young reader's perceptions' (Hughes, 1992: 147). As he explains, 'The theory behind all these stories is that a parable of this kind, absorbed by a young reader, settles into the mind's foundations as a symbolic map, a template for future psychological growth' (Hughes, 1992: 149).

In the first story, 'Why The Owl Behaves As It Does', the moral is 'glaringly simple' (Hughes, 1992: 147). Owl is a trickster figure and a tyrant. He tells the other birds that he knows 'a country where there are farms, but no farmers. You may eat when and where you please. There are no guns, no bird-scarers, no men. I will take you there if you like' (*HTWB*: 12). Hughes's owl recalls George Orwell's raven, Moses, in *Animal Farm* (1945), who also tricks the animals and claims to know of a place called 'Sugarcandy Mountain' where 'it was Sunday seven days a week, clover was in season all the year round, and lump sugar and linseed cake grew on the hedges' (Orwell, 1989: 11). Both appear as leaders in whom the other animals place their trust and both invent a place where the other creatures can live free from oppression. In the case of Hughes, Owl uses fear as a mechanism to trick the other birds and appears like a political tyrant. As Hughes explains

> the petty tyrant disguised as a political liberator brainwashes the innocent birds. He leads them into what he calls a new freedom, which is of course a prison camp of politically correct delusion where he enjoys absolute power over them all.
>
> (1992: 147)

This brings the birds who 'were so tired of their lives' (*HTWB*: 19) to want to commit suicide: 'the birds had no more interest in anything that Owl said. They only wanted to die' (*HTWB*: 19).

Like the majority of his animal poems, 'Why The Owl Behaves As It Does' exposes the child reader to the raw power and cruelty of the natural world. However, the repeated aggressive behaviour of the Owl and the violence depicted can also be read as an example of bullying and shows how by overcoming their fear the other birds regain control and defeat the Owl. Hughes provokes his readers to work out the meaning of what they read, by using common sense and wisdom, and as in his poetry he does not shelter children from violent behaviour or cruelty, but believes instead that children can cope with violence. Inevitably Owl recalls Crow, Hughes's masterpiece trickster. However, Crow's identity is far more ambivalent than Owl. Whilst Crow appears as being much more childish in his acts and is a much more complex figure, Owl is depicted as a much clearer-cut negative figure, one which can be easily recognised by young readers, despite the fact that Hughes distorts and reverses the stereotype of the wise old owl traditionally depicted in children's animal stories. The main difference between Crow and Owl is of course the fact that whilst Crow is not defeated, Owl turns out to be a failed survivor, so that when his trick fails he has no resources and at the end of the tale is revealed as a trickster, so that 'He dare come out only at night, to scrape a bare living on rats, mice and beetles' (*HTWB*: 20).

In the next story, 'How The Whale Became', we are introduced to the key figure of the creation tales, an unconventional God, and learn that he has a garden which he carefully tends. Hughes tells how the whale grew up in God's vegetable patch but was banished to the sea when he became too large. Other tales include 'How The Fox Became To Be Where It Is', which dramatises the rivalry between Slylooking, the sly fox and Foursquare, man's loyal dog, and explains how the dog became domesticated, in the same narrative line as 'How The Cat Became'. 'How The Tortoise Became' is an example of how Hughes's humanised God makes mistakes, by burning his fingers after taking 'Torto' out of his oven and dropping him. The closing tale 'How the Elephant Became' tells how Bombo, a large creature, cannot decide what to become. After the other animals laugh at him, Bombo goes to live in isolation until a great fire devastates the land where they live, bringing to mind inevitably the devastation caused by human impact on forests and the scale of deforestation linked to forest damage and biodiversity loss. However, Hughes gives his readers a happy ending, since Bombo saves the other animals and carries them to safety. Moreover, his silent strength and heroic behaviour make him 'the strongest, the cleverest, and the kindest of all animals' (*HTWB*: 92).

Amongst the other tales, 'How The Polar Bear Became' and 'How The Bee Became' also stand out as being of particular interest to ecocritics. The first is the object of Hughes's article 'The Interpretation of Parables' and begins very much in the line of 'The Emperor's New Clothes', telling of how the animals hold a beauty contest, which is almost always won by the Polar Bear, because of the special whiteness of her fur. She becomes a celebrity, besotted with vanity, and comically adored by the seals who become her faithful 'groupies', and follow her everywhere. Hughes also employs the trickster figure again here, in the character of the Peregrine Falcon, who tricks the Polar Bear into moving to the North Pole and whose role is reminiscent of that of the mirror in Snow White:

> 'I know a country', he said to Polar Bear, 'which is so clean it is even whiter than you are … There is no dirt there, no dust, no mud. You would become whiter than ever in that country. And no one lives there. You could be queen of it'.
>
> (*HTWB*: 38)

As soon as the Polar Bear (together with the seals) has gone, the rest of the animals get ready for the next beauty contest, which is won by the Brown Mouse.

From an ecocritical reading, not only is the story significant because of the obvious fact that over the last decade polar bears have become in serious danger of extinction due to global warming, but also because the tale reads as a multicultural narrative, and raises the inevitable question of why 'white' is considered as beautiful in Western culture. Furthermore, Hughes questions the real value of beauty, whilst teaching children to see through things that are artificial in order to find true values. In his article Hughes explains how 'This is a parable about the artificial self and the natural self' (1992: 148). The underlying moral of the fable is, as Hughes explains:

> When the artificial self becomes a model for behaviour, as the Polar Bear does, it becomes a tyranny, stifling true, natural life. In politics, the tyrant is removed by a coup. Within the individual, the tyranny of the false, artificial self is removed, usually, only by a catastrophic neurotic collapse and a painful psychic rebirth.
>
> (1992: 148)

Hughes's handling of the death/rebirth theme is of course closely linked to his shamanic quest. Given that shamanism for Hughes was a form of religious practice it was also a way to restore the connection between animal and human consciousness. As Hughes explains, the white polar bear lacks naturalness due to her whiteness, which is depicted as an absence of colour. In Hughes's version of the ritual the polar bear as the false, artificial self is removed so that the true self is allowed to emerge. This idea underlines much of Hughes's first book of animal fables, together with the idea that any story has the potential for use as a parable and is 'by definition a vessel of interpretations' (1992: 150).

The other tale which deserves a special mention is that of 'How The Bee Became'. In retrospective the bee was to become an important creature for both Hughes and Plath, not only in their poetry, but to the point that in 1962 Plath undertook beekeeping as a hobby. In Hughes's tale we learn that Bee is made through the precious gems and tears of a demon, and as a result must fly from flower to flower, in search of sweetness to overcome the bitter demon that runs through his veins. What is perhaps curious here is the fact that Hughes does not explicitly describe the demon as evil, but instead makes the reader feel almost sorry for him, depicting him as a sad and lonely creature, one who is jealous of God and who wants to 'make something which will be far more beautiful than any of God's creatures' (*HTWB*: 58). The irony is that he succeeds in doing so. Recalling Hughes's own remarks in his essay 'Myth and Education', the tale depicts the Christian myth in which 'Christianity is suppressing the devil and in fact suppresses imagination and suppresses vital natural life' (1970: 66). Hughes's remarks are clearly intended as a criticism of Reformed Christianity. In an ironic way then, Hughes is playing here with the Christian idea of God being a superior and 'divine' being, when, in fact, the devil tricks God, by playing to his vanity: 'I could never breathe life into it. If you had made it, it would be alive. As it is, it is beautiful, but dead' (*HTWB*: 60). Here God is gullible to flattery. Moreover, the underlying paradox of the tale is that the devil is closer to nature than God.

By confronting the idealistic Christian belief that God is benevolent and omnipotent Hughes's ideas would appear to be very much in line here with Lynn White's claim published in *Science* (1967) that the Bible is partly at fault for the ecological crisis by granting humans dominion over nature. Both characterise Christianity as being anthropocentric and argue that it desanctifies nature, and see it as being significantly responsible for our ecological crisis. In a similar way, Theodore Roszak (1973) also ascribes the roots of our ecological crisis and our alienation from nature to the Hebrew Bible, considering the heart of the crisis to be the biblical command to conquer the earth, in order to rule over it.

For Hughes, Christianity appeared as a substitute for the loss of 'the divine world' (*LTH*, 581), suppressing imagination, so that modern man has lost contact with nature. Not only is Hughes's God not the creator of the bee, but furthermore he is tricked by the demon with flattery to breathe life into the creature. The fact that the female is associated here with the demon, can be related to Hughes's own interpretation of the goddess of nature in her different manifestations as both protectively good and challengingly evil. For Hughes the goddess was the poet's threefold muse: the nurturing mother, the pure maiden or bride and the witch or hag. But above all she is manifest in nature and as a shape-shifter can possess different creatures. Following this, Hughes's parable not only mocks Christian creation and challenges it, but requires the reader to suspend scepticism and believe in his creative mythology, that is his inner world, the story within, in order to accept that even the demon can shed tears.

There is also a great deal of emotional complexity that can be found in the last paragraph which raises the question of whether or not Plath was indeed behind Hughes's thoughts at the time of writing it:

> But Bee must still go from flower to flower, seeking sweetness. The tears of the demon are still in his veins ready to make him gloomy the moment he stops drinking from the flowers. When he is angry and stings, the smart of his sting is the tear of the demon. If he has to keep that sweet, it is no wonder that he drinks sweetness until he brims over.
>
> (*HTWB*: 64)

Despite the fact that according to the archives placed at Emory, the copy of *How the Whale Became* 'included a signed manuscript note by Hughes on the first flyleaf which indicates that he wrote the poems in Benidorm in July 1956' (Shoemaker, personal communication 8 December, 2015),[18] the notebook which includes the drafts of *How the Whale Became* is not dated by Hughes. Given the fact that the book was published in November 1963 there would have been time for Hughes to have revised his original draft or even perhaps to have composed the story after Plath's death. As Hughes asserts in 'Myth and Education', referring to the power of storytelling and its importance to the child, 'Every story you tell is a whole kit of blueprints for dealing with himself and for dealing with his own imagination' (1970: 67).

Tales of the Early World *(1988)*

Hughes's second book of creation tales, *Tales of the Early World* (1988), followed a commission to write a children's story for *The Guardian*, as Hughes himself reveals in a letter to Nick Gammage, 'I wrote the Tiger story – liked it, so wrote the rest to complete a book' (*LTH*: 572).

According to Neil Philip, it is in *Tales of the Early World* (1988), where 'Hughes found a prose voice as powerful as his poetic one' (2003: 199). Certainly the language is both colloquial but richly poetic. The ten tales resemble

Hughes's earlier collection of fables but have a more developed cast of characters, together with a more imaginative and witty sense of humour, featuring a mixture of fascinating creatures, including a newt and a flea. In comparison to the first collection, the tales bear little resemblance to Kipling's fables, which read as being more simple and are far less moralistic. The fact that Hughes was not using Kipling as a model but was developing his own style makes them read as being more original. Whilst *How the Whale Became*, with the exception of God, lacks human characters, in his second book, both Man and God feature as being central to the tales. Moreover, Hughes introduces other, more complex, female characters in the form of God's Mother and Woman, Man's Wife. Hughes also draws on the dynamics of the mythic conflict from the relationship of the goddess in her different aspects, depicted largely in the characterisation of God's Mother, who is a central figure to his creation myths. Moreover, her powers appear to be far greater than those of God – who only asks for her help when he reaches the very limits of his imagination – however, when involved in the process of Creation, she takes on a very different approach to that of God. Man and Woman also both appear as central protagonists in the tales and the differences in mind and character between Man and Woman are particularly revealing, as is the relationship between Man and God.

One of the most memorable tales of the collection and perhaps the story which best examines the complexity of these relationships is 'The Guardian'. In it the characters of Man and Woman are strongly reminiscent of *Crow* (1970). Man appears as a simple creature, whilst Woman is clearly superior, much more complex and far more difficult to create: 'when he tried to breathe life into her – nothing happened. He tried again, breathing the life in very gently. She just lay there, lifeless clay' (*TEW*: 16). The tale also recalls the ending of *Ffangs*, since under Woman's heart lies a frog which was placed there by the deadly Puff-Adder, who pretends to know the cure for woman's lifelessness and offers to bring her to life. However, it is the figure of God's mother who finally brings help. The relationship between God, who lives the life of an old bachelor, and his mother, a female emanation of God who keeps the house, is particularly revealing and full of many ambiguities. God keeps her under the floor in the corner of his workshop and hires beetles to feed her. Like Hamm's parents who live in their dustbins in Beckett's *Endgame*, she lives under a heap of rubbish. She first appears banished to the corner of God's workshop, lying in oblivion, 'there she was – embedded in rubbish, like a Crab under a flat rock – his own little mother! He'd completely forgotten her!' (*TEW*: 21). Despite the fact that she is God's mother and her negative appearance being similar to that of a 'big, dry, dusty Spider' (*TEW*: 21), Hughes's goddess takes on the form of an arachnid. However, given God's futile attempts to bring woman to life, God needs the help of his mother to find a solution. In turn, God's mother requires that her son must bring her the new moon.[19] As in the narrative backstory to *Crow* (1970), God's most ambitious project is to create Man and Woman. Using the new moon, blood and her magic, God's mother creates Tiger, 'the Protector of the Human Child' (*TEW*: 26) and Baby, who in turn bring Woman to life.

Amongst the many Hughes critics who have discussed the central significance of Hughes's female goddess in her different representations, such as Keith Sagar (2006), and especially Ann Skea (1994), Usha V. T. (1998) has focused on the issue of the representation of femininity in Hughes's poetry and on 'the earthly representatives of the Moon Goddess, the fertility principle' (Usha, 1998: 202). In line with Usha, by associating the goddess with the moon, it could also be interpreted here as a fertility figure, one that recalls Hughes's own lost muse/goddess, Plath. It should also be remembered that it was Hughes who introduced Plath to Graves, following which, Hughes's own interest in shamanism and the goddess had also influenced Plath's own writing.[20]

Like Graves's goddess, Hughes's goddess, as we have seen, represents the complexity of creative-destructive forces in nature and can be read as the underlying ecological essence of Hughes's work. The main difference between Graves's and Hughes's goddess would appear to be the fact that whilst both are concerned with the creative and destructive driving forces which find their most natural expression in the natural world, they go about it in different ways. Graves seeks to control the energy, whilst Hughes takes a Jungian approach, describing what is needed to establish a balance between the male/female forces, so that his entire poetic quest seeks to reconnect alienated Western Christian culture with nature and the inner self. Hughes's view is more complicated and sophisticated than that of Graves. For Hughes the goddess was manifest above all in the natural World, but she was also a symbol of the feminine aspect of the male psyche, or what Jung termed as anima. However, as Janne Stigen Drangsholt points out, the merging of Jung and Graves's gendered metaphors can raise a lot of questions, especially for ecofeminists. As she puts it 'The problem with associating these goddess figures with femininity, then, is that they can end up reinforcing certain essentialist ideas of what it means to be female' (Drangsholt, 2015: 86).

Numerous critics have commented on the ambiguity associated with the female figures presented in Hughes's poetry (Anderson, 1994: 91; Bentley, 1998: 97 and Brain, 2011: 94–106). Paul Bentley notes that Hughes's representation of the feminine in his poetry come dangerously close to damaging essentialist stereotypes (1998: 97). Whilst agreeing, to an extent, that the notions of the feminine enacted by Hughes in his adult poetry do attract charges of essentialism, it would seem that throughout his children's work, the poet's attempt to restore the rift between dualities, including the masculine and feminine, is more successful given the genre in which it is written. In the case of Hughes's creation tales, this can be seen in the playful way in which Hughes depicts both the male and female figures and explores these tensions. Moreover, despite the fact that God is male, it is important to remember that it is often God's mother who saves the day. In a similar way, the character of Woman is far more of a psychologically complex and developed figure than her male counterpart.

In *The White Goddess* (1948), Graves claimed that the goddess was forced underground by the rise of patriarchy. The fertility figure of the moon, as a representation of the goddess, is used by Graves as a demonstration of a mythical process of recovery and growth, one that can only be enacted by woman. Such ideas would, of course, not be shared by ecofeminists. In contrast with Graves's 'silent Muse' (Graves, 1966: 447), Hughes's 'muse', in his creation tales, is far from voiceless. Since the rejection of the female results in mankind's alienation from nature, from an ecofeminist reading, Hughes deals with this in his creation tales in a holistic way, in his attempt to reconcile the underlying conflicts depicted in the relationship between Man and Woman and the creative methods of God and his mother. The suppressed female energies appear at first in the figure of God's mother, and are later revealed as being the alienated female energies of Nature. He enacts his myth of the goddess of Nature in the tale of 'The Guardian', by symbolically displacing the mother figure 'underground' in God's workshop. God is unable to complete his female creation and needs the help of the goddess, who appears here in the maternal form of an old hag-like arachnid figure, to find a solution to his creation of Adam and Eve: a solution that comes in the form of a baby child. In this way, the new moon symbolically is the white goddess of fertility and purification. The baby, as a representation of birth and renewal, is something that can only be endowed by the female goddess, and not by God himself.

Despite the fact that Hughes's God may blame others for his failures, or lack fallibility at times, and is often not even aware of what is going on, the tales do not depict a negative view of God. Hughes also makes sure that God is depicted as being an empathetic and benevolent creator. The rest of the fables narrate how God continued to craft and create more creatures of the world: by accidentally breathing life into a miniature trunk made out of the remains of elephant clay God creates the earthworm; when he tries to make a dancer he first produces Cat and later Mouse, hence their rivalry; in 'How Sparrow Saved the Birds' we learn of Sparrow's heroic battle against the bird-swallowing Black Hole. On other occasions the tales are strongly reminiscent of *Crow* (1970), such as in 'The Playmate'. Hughes attributes reductive gender roles to Man and Woman in so far as their responsibilities and chores. However, given that Woman is more complex than Man, she often gets bored with Man's ignorance and her simple life of weaving baskets and making pots. After coming up with an uncontrollable Gorilla as a playmate, which ends in disaster, it is Woman who helps God imagine the perfect playmate for her, one that is 'Beautiful – like the sea' and 'exciting' (*TEW*: 79) – like the Horse.

Another of the most outstanding tales is Hughes's own Crow-like adaptation in 'The Shawl of the Beauty of the World', in which we learn how God created the peacock: 'it was a funny-looking object. Just like a Turkey all plucked and ready for the oven. Even its head was pink and bare. And it had no legs' (*TEW*: 85). The tale also introduces two new characters: Boy and Girl. They steal into God's workshop and interfere with one of his unfinished works – a featherless bird with no head or feet, called Poor Thing. Playful and childish, they behave much like the infantile trickster character of Crow, and mistakenly give him Demon feet and the wrong voice. In a similar way to 'The Guardian', it is God's mother who takes pity and saves the creature, by lending the strange bird a beautiful shawl, a present from God, thus creating the first peacock.

In Hughes's creation tales nearly all the creatures share a similar preoccupation: how to deal with the challenge of becoming. They are all displaced creatures seeking the divine as part of God's creation, in a world where God enacts their Genesis, by subverting the narrative and transforming them from nothingness and endowing them with life. Despite the fact that the tales are playfully subversive they also give insight to Hughes's empathy towards nature and his concern for the natural world, creating a sense of wonder and respect in the reader. God's blunders and fallibility also make him much closer to human beings. In this way Hughes sensitises his readers and provokes compassion in them, compelling the young child – and future adult – to take their lives in their own hands and act for themselves.

The Dreamfighter *(1995)*

The third and final collection of Hughes's creation tales, *The Dreamfighter and Other Creation Tales* (1995), was dedicated to Carol Hughes. It includes 11 new tales in which we learn more about the powers behind Hughes's creative universe and which take us into an extended cosmos: Heaven, the Earth, the Moon and the Underworld. The complete collection of Hughes's creation tales were also collected in the *Collected Creation Tales*, which was also titled *The Dreamfighter and Other Creation Tales* (2003). Like *Tales of the Early World* (1988), the collection revisits and extends the first collection of fables, with a stronger element of irony and religious satire familiarly reminiscent of his *Crow* saga (1970). God continues to create all kinds of creatures but demons keep reappearing. After all, as Hughes reminds us in the finale of his epic poem sequence in *Cave Birds* (1978) 'at the end of the ritual, up comes the goblin' (*CP*: 440).

In the first tale 'Goku', Genesis takes place again. The tale narrates how Man and Woman were created by the Angels and Demons, and tells how 'Suddenly God had a brainwave. If Goka has a baby, he thought, she will calm down. She'll become sensible. And Goku too, he will become serious. Fathers become serious' (*DF*: 200–201).

According to Jonathan Bate the story offers a glimpse of Hughes's life, his fatherhood, 'transposed into mythic narrative' (2015: 495). However, Bate's comment appears somewhat flippant since he fails to go into any depth as to whether Hughes thought Plath would calm down after having children or whether Hughes thought he would. In the narrative, like sadly in Hughes's flawed life, things do not, however, turn out as planned: 'And before God came back, Man and Woman had two children, Boy and Girl – but their heads were half turnip, just as they are to this day' (*DF*: 199). Given the fact that 'Goku' was first published in 1995, it would seem that the image of the turnip head can be read here as a caution from Hughes concerning the corrosive effects of the modern technological world, and an education system which places its trust in scientific certitude rather than imagination and creativity. Hughes's cruel and sardonic sense of humour becomes more subversive and somewhat sardonic, similar to Roald Dahl's black humour, used to ironise but in a playful way.

In 'The Dreamfighter', the second tale that gives its name to the collection, God advertises for a bodyguard to defend him from nightmares and Space Beings. 'The Mutant' becomes God's protector, however when his ego grows, like him, in size to teach him a lesson God turns the Giant into an ant. Along with the other creatures and characters, we meet a Poltergeist who lives in God's workshop in a candlestick and delights in playing tricks on God. Some of his creations are failures, like 'Camel':

> Camel was a mistake. He was simply made wrong. He stood in God's backyard, where the different kinds of clay and the fuel for the kilns lay in heaps, and he knew he was wrong. And he felt wrong all over.
>
> (*DF*: 275)

Previous characters like the Demon, also appear again, and trick the camel by taking on the shape of an Angel. Most of the tales have underlying morals, such as in 'The Grizzly Bear and the Human Child', when God gives Woman a toy so that 'While Man was away in the forest, collecting food, she had to have something to play with' (*DF*: 288). Given her fancy for jewels he decides to give her a 'living jewel' (*DF*: 288) and invents Snake. However, Snake turns out to be far more than a simple jewel and thanks to his power drives away Woman's other playmates and even learns to take on the form of Woman herself. When Woman asks Man to ask God for a baby child, Snake decides to trick them and disguised as Woman goes to visit God, claiming that she wants a grizzly bear instead of a baby. Despite seducing God, who in turn sends Man and Woman a grizzly bear, God finally amends his mistake by leading Grizzly Bear to the forest and brings them a baby.

'The Last of the Dinosaurs' tells of a God who failed to make things better. We learn how God got carried away in the early days and filled the Earth with dinosaurs. After noticing that he had 'filled the Earth with monsters' (*DF*: 305) and that the world was no longer a safe place for the rest of the creatures, God then sets out to wipe them out, right up until the last one, the Pterodactyl, which he turns into a heron.

In the last tale 'The Secret of Man's Wife' we learn about the bond between Moon, Woman and Hughes's favourite totem figure and mythic animal, Fox. However the tale also offers an interesting psychoanalytical reading about the female 'time of the month' syndrome,[21] in which Man's Wife yearns to escape whilst he believes she is meeting a wolf in the woods. In the narrative Man asks God to capture the wolf that is making her act strange and tame it as a pet in order to make his wife normal again. The reversal of stereotypes depicted by Hughes, and the critique of male behaviour, is of course, intended to be ironic to highlight Hughes's point that men do not understand women. In the tale it is God's mother, who despite having less protagonism than in the previous collection, comes to help yet again. God's mother, Hughes's own enactment of the goddess, is much more than a female emanation of God, and clearly a more powerful being. Woman has been possessed by a Demon, so she performs a particularly ferocious exorcism in order for her good Angel to be liberated, in which she is put 'head-first into the furnace' (*DF*: 318), and takes on the form of a fox. It seems unlikely that it is merely coincidence that Hughes's favourite and most important totem symbol, the fox, appears here at the very end of the last tale of his creation tales to close the collection, illustrating the many interconnections between Hughes's animal motifs in his different writings.

Read as a collection, Hughes's creation tales also disclose some revealing and rather puzzling facts. Hughes's choice of employing the creation myth as a narrative technique allowed him to rewrite characters and offer parallel readings though different accounts of the same story. As Kazzer points out in his essay 'Family relations – Traces of a Cosmology in Ted Hughes's Creation Tales', this reveals 'certain inconsistencies in them as regards some of the characters and/or their creation' (2005: 2). These inconsistencies refer mainly to conflicting narratives regarding the creation of Man and Woman, the human child or other creatures, such as the figure of the snake. However, despite the apparent contradiction, there would appear to be a reason behind Hughes's choice of rewriting inconsistent and even ambiguous accounts. As a storyteller this gave him the freedom to adapt a certain tale to particular circumstances, situation or audience and the possibility to retell the same tale.[22]

Indeed, the tales would appear to portray certain dualities such as the Demon/Angel or split creatures that exist side by side with other creations. It would appear that these partly conflicting accounts make Hughes's God and creatures resemble certain characteristics of Gnosticism. Although Hughes was not a declared Gnostic he was deeply influenced by early animist and pagan religions, and fascinated with the occult. So by offering different accounts of the same narrative, by presenting conflicting tales about the creation of different characters, Hughes was trying to stimulate the reader's imagination as well as enacting his own creative myth.

Reading *The Iron Man* and *The Iron Woman* as healing myths

Hughes's children's writing is dedicated to the ongoing process of recovery and renewal of the imagination of the new child, extending his underlying idea that children have the chance to recover the lost sense of awareness and wonder. Hughes's love of the natural world and in particular the therapeutic function of healing that he ascribes to nature, is part of a major professional project which the poet embraced throughout his entire career. As Sagar has argued, the vital function of literature is to heal the wound that culture has inflicted upon nature. Following this, both *The Iron Man* (1968), and its sequel *The Iron Woman* (1993), read as Hughes's clearest depiction of ecological healing and draw strongly on Rachel Carson's idea that harm to nature is caused by humankind.

In them, Hughes advocates his concern, speaking on behalf of nature, and warns of the dangers of toxic waste and river pollution. Linked with his concern about the water quality in the nation's rivers, both stories engage directly with our ecological crisis, alerting not only to the dangers of manipulating natural resources and destroying the environment, but also serving as an indicator of human water pollution. Although both stories run similar themes, their main difference lies in the intended audience of his creative work. Whilst *The Iron Man* (1968) serves as a healing myth on a more personal sphere, for Hughes himself, and is addressed to his children, its sequel, *The Iron Woman* (1993), written 25 years later, draws more on the centrality of Carson's sense of wonder, and can be read as a more convincing healing myth and a redemptive story for a social ecology.

The Iron Man *(1968)*

The influence that folklore and anthropology had on Hughes's career can be seen throughout Hughes's storytelling as a cultural project. For Hughes, as a storyteller, myths are narratives which tell the exploits of heroes, supernatural events and distant universes. They act as a way to negotiate between the outer world and our inner life. In Hughes's creative universe, myth and ritual go together. With them the world, however chaotic it may seem, is able to take on meaning and form. This is, at least, what Hughes tells us in his essay 'Myth and Education' (1970), in which he explains that *The Iron Man: A Children's Story in Five Nights* (1968),[23] is a story about a boy, Hogarth, who takes control in the face of the threat of the Iron Man and ultimately restores peace on earth. However, given the context in which it was written and from a Jungian reading, it can also be read on a much more personal level, as a healing myth and a story of redemption. Hughes himself outlines the moral purpose of his storytelling and analyses the meaning of myths in his essay where he expresses the idea that some kinds of imaginative stories can have a curative effect and that as adults 'they are hospitals where we heal, where our imaginations are healed' (1970: 67).

Hughes wrote *The Iron Man: A Children's Story in Five Nights* (1968) as a bedtime story for his own children, as a kind of healing and coping mechanism following the death of their mother. Bearing in mind the traumatic context in which it was written, it reads on a more personal level, as a healing myth and as a story of redemption. In 1968, when the book was first published, Hughes's own children were still very young. His daughter Frieda would have been eight, Nicholas six and Shura was just four. Their names[24] on the dedication pages in various editions of *The Iron Man*[25] also make visible Hughes's reasons for wanting to create such a powerful creative story.

Hughes focused on the reading of the story as a healing mechanism, both for himself and for his young children. Speaking about the musical quality of his story, in an interview, Hughes recalls how he used to read the story to his children to the beat of a large African bongo drum, until his daughter Frieda asked him to stop using it and told him that they didn't like the drum and that it frightened them.[26] Certainly the story has a strong underlying rhythm and reads best aloud. As Paul has pointed out, 'It has the meticulous, ritual slowness that occurs with frequent, intimate domestic retellings' (2000: 219).

However, as Hughes claims, *The Iron Man* was not just a story for his own children, but one that he had written 'as a blueprint imaginative strategy for dealing with a neurosis.... It is a story intended to cure the mentally sick, &, to put people in contact with their real nature' (*LTH*: 284). So Hughes explains the Iron Man, as an imaginative blueprint not just for his own children, but one available for all to access in order to harness the strength of threatening monsters, which he does by incorporating them into the story and then turning them into good.

Halfway between a modern fairy tale and science-fiction myth, it describes the mysterious arrival of a giant 'metal man' who comes from out of the sea, and after falling off a cliff manages to reassemble himself. Because he only eats metal, he soon becomes a problem for the local farmers, feasting on their tractors and raining destruction on the countryside by attacking their farm equipment. In the narrative he represents the machine age, twentieth-century technology, controlled and at the service of mankind. Hughes personifies the wake of the ongoing environmental conflict between the age of machines and technology, and the destruction of the environment, through human-induced changes to nature. *The Iron Man* is Hughes's version of the archetypal heroic figure, one who offers himself for sacrifice in order to restore life on earth. However *The Iron Man* should not only be read as an archetypal survivor's tale, but also as a visionary story, one about restoring life to earth, about being reconnected and made whole.[27]

The Iron Man is also a story of friendship, betrayal and reconciliation, one that takes place between Hogarth, the little boy,[28] and the Iron Man. When the farmers decide to dig a giant trap, Hogarth helps them to capture and bury him. However, the Iron Man has a primitive power and magic about him, and after being buried he rises again. After reconciling, Hogarth decides to help the giant 'metal man' by leading him to the scrap yard where he can feast on the mountains of old cars and second-hand machines rather than on the tractors belonging to the local farmers.

In the second half of the story Hughes introduces a new threat, in the form of a giant space-being called the 'Space-Bat-Angel-Dragon' which threatens the earth. The Iron Man functions here as a counterpart to the George and the Dragon story. By creating something that was the opposite to the legend of George and the Dragon Hughes was attempting to create 'a sort of imaginative blueprint' to counter 'the symbolic story of creating a neurosis' (1970: 66). This is exemplified in the story when the Iron Man defends the world from the strange space creature, which is then exiled from earth and finds its place in the universe by flying. In this way the movement of its wings restarts the music of the spheres, which has a calming effect on the people of the earth, and brings an end to all the conflicts and wars over the globe:

> The strange soft eerie space music began to alter all the people of the world. They stopped making weapons. The countries began to think how they could live pleasantly alongside each other, rather than how to get rid of each other.
>
> (*TIM*: 62)

The way in which Hughes describes the music here suggests a creative displacement, a reverse of St George and the Dragon. The music of the spheres is about turning a potential threat, that of the machine age personified in *The Iron Man*, into something positive so that by finding a place for the other, it can have a life of its own, a utility, and so be meaningful to society.

As in his creation tales, Hughes is using the 'parable' here, by definition, as 'a vessel of interpretations' (1992: 150), one which serves not only as 'a template for future psychological growth' (1992: 149), but one which enacts the myth of harmony and peace, so that as a visionary story, it restores life to earth. So from an ecocritical viewpoint, *The Iron Man* is a fable about the modern world, about human ignorance, or rather that of adults since the world is saved thanks to young Hogarth, who faces his fears. In line with Hughes own healing quest, culture (in the form of the Iron Man representing technology) and nature are brought together again in harmony at the end of the story.

However, perhaps the main reason why *The Iron Man* reads as a more personal healing myth for Hughes, in comparison to *The Iron Woman* (1993) which has a more holistic ecological message, is because Hughes attempts, in an allegorical sense, to guide his children through the healing process of the loss of their mother by creating the Iron Man as a heroic figure and using the music of the spheres as a healing archetypal image to transmit a sense of harmony and hope. In this sense it reads as being as a redemptive blueprint story, one that talks, in line with Jung, about being reconnected to our inner self, about being restored and being made whole.

The Iron Woman *(1993)*

Whilst *The Iron Man* serves as a healing myth in a more personal sphere, *The Iron Woman* was written as an intervention on behalf of water quality and public health, and has a much stronger and active environmental agenda. The sequel draws on the centrality of Carson's *Silent Spring* (1962) as well as *The Sense of Wonder* (1965), and can be read as a healing myth and a redemptive story for a society that has cut itself off from 'being human' and from being part of the larger web of life.[29]

The book begins with Hughes's critique of a birdwatcher who having spent an entire day staring through his binoculars, fails to notice that that the bird he is watching is actually dead. The birdwatcher would appear to represent the detachment of human culture from nature and recalls Master, the central character in Hughes's children's play 'The Tiger's Bones' (1970), who mistakes a lump of mould in his telescope for a meteor that is about the hit the earth. The fact that both characters see the world through a narrow scientific lens prevents them from seeing what really lies before them.

The tale takes on a similar plot to that of *The Iron Man*, with the Iron Woman appearing not from the sea but instead from a black polluted marsh. When the birdwatcher sees the Iron Woman rise out of the swamp he runs away in fear. As in *The Iron Man*, Hughes chooses a child as the first to identify and empathise with the problem in order to bring about a resolution. The Iron Woman speaks to Lucy, the young female protagonist, in the name of the creatures that live in the river, vowing to destroy the factories and those who have dumped toxins and are polluting the river and marshlands. It is the young girl, Lucy, who is the first to realise that something is wrong with the fish in the river as she observes a silent writhing eel:

> Seeing its peculiar head bobbing up like that, and its little mouth opening, she had felt a painful twist somewhere in her middle. She had wanted to scoop the eel up and help it. It needed help. Something was wrong with it. (*TIW*: 2)

She decides to work with the Iron Woman to change the ways of mankind and calls upon the help of Hogarth and the Iron Man. As she tells the factory manager, together with Hogarth: 'Your factory has poisoned the river. It's killed all the fish … You're poisoning all the creatures and you're also poisoning me' (*TIW*: 41). However, when the children's pleas are ignored, the Iron Woman decides to teach the men at the factory a lesson by turning them into fish. In order to survive they have to be taken to the water by the women, and so experience in this way the pain of the poisoned creatures that live in the polluted river.

If we look at the way in which the Iron Woman directs her revenge on a male community (that of the workers at the waste disposal factory) for polluting water resources, an ecofeminist reading of the novel can offer insight into the close relation that Hughes depicts between the Iron Woman and nature, as well as looking at how the duality of nature/culture is depicted. For Hughes the image of the female as nature plays an important symbolic role in constructing his view of human identity. Furthermore, the novel explores how Hughes depicts the male/female duality: whilst the male factory workers and greedy managers are responsible for poisoning the river, the women are depicted as the 'goodies'. Blake Morrison even put forth the claim that *had* the book been written by a woman, it would have been called feminist.

In the words of Roberts, the Iron Woman is 'perhaps Hughes's most direct representation of the Goddess' (2006: 177), whilst according to Gifford, *The Iron Woman* (1993) offers 'one of Hughes's most powerful and educational interventions on behalf of water quality and public health' (Gifford, 2011: 90). Along these lines, Yvonne Reddick has recently called *The Iron Woman* 'an environmental fable' (2017: 10) and claims that it 'goes some way towards deconstructing the stereotype perpetuated in 1970s counterculture and early

1990s ecofeminism, of woman as "close to the Earth" and man as technologically superior' (2017: 10).

Given that she rises from the polluted river, the Iron Woman can be conceived as Hughes's river-goddess, symbolising for Hughes the goddess of Nature in her clearest form, crossing the borders between human and nature, and mediating between both. The fact that she transcends the boundaries between the artificial and the natural makes the narrative in the words of Zoe Jacques 'a cybernetic complication' (2015: 197). However drawing on ecofeminist Donna Haraway's term 'natureculture'[30] as a single word for the dissolved duality, it could also be seen here as unifying. Perhaps as Nina Lykke suggests, if we compare the cyborg and the goddess they have more in common than first meets the eye (Lykke and Braidotti, 1996):

> Both are, so to speak, designed to transgress the borders between human and non-human. Both challenge the ways in which the modern scientific world-view is rooted in a long tradition that casts the non-human in the role of a mere object and exploitable resource for the human.
>
> (Hovender *et al.*, 2000: 82)

Lykke, together with other more spiritually-orientated ecofeminists, would appear to argue that a feminist reclaiming of the great goddess of nature may help to redirect society. In line with Hughes's own healing quest, the Iron Woman, as half goddess/half cyborg, puts forward a similar redemptive idea, and reads as a potential healer of broken bonds between humans and nature.

In an interview with Morrison, Hughes describes what is needed to establish the male/female balance and explains why it is necessary:

> Men are the perpetrators of the crime but not necessarily destructive.... Men and women are two parts of one system, not two separate systems. I wanted to reach that balance and make sure that nobody was blamed for things going wrong.
>
> (1999: 168)

Hughes's arguments would appear to approach those of Val Plumwood, who argues that all systems of dualisms such as male/female, culture/nature stem from the Western conception of the order of the world. Whereas for Hughes, who would appear to hold a more historical position, the male and female complement each other in a holistic sense, Plumwood – along with other ecofeminists – would consider such a view as being conservative and undermining. However, perhaps the real problem lies in the definition of human, by what is seen as distinguishing humans *from* nature rather than including what humans share *with* nature. Whilst Plumwood, who rejects labels such as 'essentialism', argues that whilst we cannot simply ignore the woman/nature link it should be avoided, as the idea of a 'female' nature is one that has been systematically degraded; for Hughes, the image of the female as nature plays an important symbolic role in constructing his view of human identity and one that becomes complete in *The Iron Woman*. Notwithstanding, Hughes's position is rather awkward, and remains somewhat problematic from a historical and cultural position, since in the novel he aligns women, as the maternal carers, with nature, whilst the men are responsible for the polluting the river, representing a male-dominated capitalism.

Although it was published in 1993, Hughes had already begun writing *The Iron Woman* as early as 1988, at the same time as he was writing *Shakespeare and the Goddess of Complete Being* (1992). However, after completing the first two chapters (Sagar, 2012: 234), he turned back to *Shakespeare and the Goddess of Complete Being*, which was finally completed and published in 1992, after which he returned and finished off *The Iron Woman* the same year. Speaking to Blake Morrison about this Hughes claimed that:

> I began writing the story in the mid-80s, and at one point I was scared by it and had to back off. The image of that scream in particular alarmed me. I wasn't sure what I was pushing myself into, so I left it alone for a bit and turned to Shakespeare instead while I got used to it.
>
> (1999: 160)

Undoubtedly the most memorable thing about the Iron Woman is her contagious Macbeth-like scream – the cry of the marsh – which recalls Hughes's own haunting voice of pain:

> 'It is the cry of the insects, the leeches, the worms, the shrimps, the water skeeters, the beetles, the bream, the perch, the carp, the pike, the eels'. 'They're crying', whispered Lucy. 'The cry of the ditches and the ponds', the voice went on. 'Of the frogs, the toads, the newts. The cry of the rivers and the lakes. Of all the creatures under the water, on top of the water, and all that go between'.
>
> (*TIW*: 20)

The Iron Woman's cry can be seen here as a response to Hughes's poetic self, or to what Morrison called Hughes's 'voice of pain' (*LTH*: 458). Referring to the tragedy Hughes endured in his life Morrison even goes as far as suggesting the possibility of linking 'the central image of the scream to the noise Hughes has endured in his own life for more than twenty years' (1999: 169).

In the novel, when things turn into a national disaster and get out of hand, the children beg the Iron Woman to stop, but she insists that the people of the earth still haven't learned their lesson and that there must be a 'deep change' (*TIW*: 85). Hughes himself advocated the need for that same deep cultural change first in his urgency to do something facing the environmental damage of his local rivers, and later to transform what he called 'the soul-state of our civilization' (*WP*: 130). As Richard Kerridge puts it, Hughes's view is that a 'deep cultural change is required to rescue Westernized humanity from terminal ecological catastrophe and also from that insatiable exiled quest' (2015: 187). In the novel, Hughes's urgency for that deep change comes in the form of a healing miracle to save nature and humankind. Meanwhile the men emerge from the water with the screaming ringing in their ears and their hair turned white, as if bleached or aged. Mysterious yellow webs appear to have grown overnight covering the rubbish. When mixed with water this became a magic fuel, that is, an environmentally friendly source of energy, one that could be dissolved in water and was non-toxic in rivers. When Lucy and Hogarth ask the Iron Woman where the yellow webs come from she tells them the answer, that only a 'Big, deep fright' can generate a 'Big, deep change' (*WP*: 85).

The novel's closing picture is one of hope, when the Iron Woman puts a garland of foxgloves round the neck of the Iron Man, and one of snowdrops around her own.[31] Hughes's choice of a foxglove, universally accepted for treatment of heart disease, is particularly interesting, bearing in mind that it's a plant that heals the heart. It reads here as a symbol of reconciliation and healing, appearing at the beginning of the novel when the Iron Woman visits Lucy for the first time and gives her both flowers and again at the end.

The redemptive aspect underlying the story also reads as being closely linked to Hughes's own concern to heal the rift between culture/nature. Throughout his career Hughes was well aware that the global environmental crisis that we were facing and the effects that the widespread use of pesticides and pollution had on the environment was largely of our own making. As he said:

> Most people I talk to seem to defend or rationalize the pollution of water. They think you're defending fish or insects or flowers.... It isn't a problem of looking after the birds and bees, but of how to ferry human beings through the next generation.
>
> (Morrison, 1999: 165)

The other major problem was, of course, how to get this message across to the rest of society. In a letter Hughes wrote to his editor, who was at the time Christopher Reid, at Faber, he made the following proposal referring to *The Iron Woman*: 'We could send John Major a gold-backed copy. Present all the chieftains with one maybe ... And all the cabinet' (Gifford, 2009: 25).[32]

So *The Iron Woman* reads not only as a redemptive healing myth, as the poet's personal reconciliation with that nagging 'voice of pain', but as his clearest depiction of ecological healing. Hughes was enacting his greatest healing myth and through redemption, in the form of saving not only nature but by reconnecting culture (society) to its roots (nature) he was completing the final part of his own healing quest. In his healing myth, and following Carson's steps, Hughes points to modern capitalism as the culprit of human exploitation of the earth and the cause behind humanity's environmental crisis. Hughes had faith in the power of nature to restore itself and believed that that not only can nature be saved but that humanity *can* change if it is willing to rethink its attitudes towards its place in the web of life. However, unlike Carson's cautionary tale, *The Iron Woman* is optimistic (perhaps too optimistic if we look at how slowly society is responding to global warming).

Hughes's portrayal of human's self-inflicted destructiveness, written as an intervention on behalf of water quality and public health, alerts us not only to the dangers of manipulating natural resources and destroying the environment, but also serves as an indicator of human water pollution. By explaining the problem Hughes seeks a solution and compels in the young reader an ethic of stewardship. Hughes not only includes the concern for human species in relation with nature, but his call for a holistic ecological sensibility brings him in line with Murray Bookchin's concept of 'social ecology' (1982), in his idea that ecological problems stem from social problems. In his narrative, Hughes's critique of humanity – in this case a male-dominated capitalist society – suggests, like Bookchin, that ecological problems, such as the river pollution depicted as both a local and global threat, also stem from social problems. By showing the implications for human responsibility Hughes successfully explores how human alienation from nature can be understood and through the figure of the Iron Woman manages to reconcile the ecological sense of 'healing' by bridging the duality of culture/nature together. Hughes is, of course, clearly speaking to an adult audience in his critique of human alienation and the detachment of modern society from nature that underlie both healing myths. Both texts read as unmistakable examples of the disconnected adult/child audience that he was addressing, and draw on the importance of breaking down barriers. Moreover, both novels are examples of how children's fiction can successfully engage with the entangled relationship between society, technology and nature and surpass the boundaries that separate nature from culture.

Hughes aspired throughout his life to teach children to respect the natural world through a positive connection to nature. Hughes's children's writing is dedicated to this ongoing process of recovery and renewal of the imagination of the new child, extending his underlying idea that children have the chance to recover the lost sense of awareness and wonder. His writing not only questions where we are today in our understanding about the sense of wonder in young children but urges us as adults to use all our senses and emotions to explore nature in order to reconnect our inner selves with the potential of our imaginative consciousness. However, despite our growing environmental awareness, judging from where we stand today there is still much to achieve. In view of a future that is anything but simple, it is thus that Hughes places the enormous responsibility of saving humankind indirectly on the child's shoulders, for as Wordsworth said, 'the child is father of the man' (1984: 246).

Notes

1 Yvonne Reddick's recent book *Ted Hughes: Environmentalist and Ecopoet* (2017) offers an innovative analysis of Hughes's writing by looking at how his ecopoetics reflects his own environmental awareness whilst paying attention to the historical and political context.

2 When Hughes's friend Jim Downer showed him the drawings and rhymes he had written for a children's book *Timmy the Tug* in 1952, Hughes replied that he like the drawings but would write the verses himself. More than 50 years later Carol Hughes found the manuscript and returned it to Downer. It was finally published in 2009 by Thames and Hudson.

3 The hasty ending of *Ffangs* resembles that of *Crow* (1970), which stops at the point when Crow crosses the river – just as he is about to meet and be united with a beautiful maiden, whom the hag turns into. It could be easily argued that Crow's unhappy ending was due to Hughes's own tragic personal experience with Wevill at the time. This would also validate a redemptive reading to *Ffangs*, as Hughes's own personal attempt to give a 'happy ending' to the story, which ironically he fails to do.

4 *Falcon Yard*, is the title of an unpublished novel which Plath worked on between 1957 and 1959, which fictionalises Plath and Hughes's life and the party in Cambridge where Plath met and famously bit Hughes. The book was thought to have gone missing, but notes from the book were discovered amongst archives at Emory University in 1999 and can be found in MSS 644, Box 139, Robert W. Woodruff Library, Emory University, Atlanta, Georgia, USA.

5 For an ecofeminist reading of the moon goddess which appears in Hughes's moon poems see chapter five: *The Earth Owl and Other Moon People* (1963), which also analyses the moon as a symbol in Plath's poetry.

6 See Armitage's (2010) unpublished thesis.

7 Morrison, 'Man of Mettle' was first published in the *Independent on Sunday*, 5 September 1993, Review pp. 32–34.

8 The proofs of a different version to *Ffangs* date from 21 June 1979, and can be found amongst notes that lie in a 'box of endless tales' in the Hughes archives of the British Library, together with a loose sheet undated which appears to be a draft of 'Ffangs 2'. The draft talks about resurrection, immortal vampires and what is perhaps more curious is the fact that on the reverse sheet Hughes mentions the Iron Woman: 'Iron Woman? – not final story – mention Ffangs, the snake, dying forest, Whales are dying' (Ted Hughes Papers, Add MS 88918/8/5, The British Library, London).

9 Nicholas had just finished a BA degree in zoology at Oxford and would dedicate his life to the study of fish ecology. At the time he was working in Fairbanks, Alaska, as a research assistant at the Alaska Cooperative Fish and Wildlife Research Unit.

10 All three books of creation tales were collected in a book which is also titled *The Dreamfighter and Other Creation Tales*, published by Faber and Faber, 2003. Hughes also wrote a story about a 'stray' as a contribution that appears in Michael Morpurgo's collection of stories *Muck and Magic* (1995).

11 Amongst the files Hughes deposited in the British Library Archive, Add MS88918/8/2, Box 12 reads 'file containing notes and drafts for stories of various kinds, partly written and unpublished' in Hughes's handwriting. The named file contains 25 school exercise notebooks each with its title inscribed containing materials in various stages. The titles include 'Boa Constrictor', 'Cockroach', 'Garden Spider', 'Dragonfly', 'Rhino', 'Mosquito', 'Yeti', 'Snow Leopard' and 'Pike'. Despite the fact that many of these tales sadly never saw the light, it is interesting to note that most of the original drafts of Hughes's creation tales flow from the start without hardly any alterations to the handwritten version when compared with the final published copy.

12 Hughes reviewed Golding's second published novel, *The Inheritors*, from which he also read at his memorial service. Hughes's essay entitled 'Baboons and Neanderthals: A Rereading of the Inheritors' in John Carey (ed.), *William Golding: The Man and His Books: A Tribute on His 75th Birthday* (London: Faber and Faber, 1986), pp. 161–168, is, more than a rereading of Golding, an exploration on how human beings are an evolutionary error. In it Hughes develops his theory that the myth of the Fall can be traced anthropologically.

13 Amongst the simplest of myth or creation stories are the why stories, or *pourquoi* tales, that explain the existence of why something is and how it came to be.

14 The *Jatakas* are amongst the earliest Buddhist literature, dating as early as the fourth century BC, concerning the previous lives of Gautama Buddha, in both human and animal form. For more information on their influence in Kipling see Peter L. Caracciolo, 'Buddhist Teaching Stories and Their Influence on Conrad, Wells and Kipling: The Reception of the Jataka and Allied Forms in Victorian Culture', *The Conradian*, 11:1 (8 May, 1986), pp. 24–34.

15 Untitled notes by Ted Hughes on the universal goddess. MSS 644, Box 115, Robert W. Woodruff Library, Emory University, Atlanta, Georgia, USA.

16 Hughes appears to prefer the term 'parable' to 'fable' to refer to his creation tales. Whilst parables generally feature human characters and are used allegorically to illustrate a moral attitude or religious principle, fables use animals and inanimate objects as characters, which are anthropomorphised to illustrate general truths about human nature. In the case of the parable if the story is told in terms of animals, they are never personified, but remain strictly themselves. In the case of a fable, the lesson is not allegorical as each animal is not necessarily a symbol for something else. Instead the reader learns a moral lesson as an example of what one should or should not do. So, strictly speaking, these are fables.

17 The story in question was the story of 'How the Polar Bear Became'.

18 My thanks to Kathy Shoemaker, Reference Coordinator, Research Services at Stuart A. Rose Manuscript, Archives, & Rare Book Library, Robert W. Woodruff Library, Emory University, Atlanta, for answering my queries via email regarding this.

19 The image of the goddess related to the moon is a recurrent one in Hughes's writing. In *Ffangs*, the female's association with the snake in the story is combined with that of the moon in the character of Selena, who is found on the moon with a snake in her mouth whose bite reveals the true character of people.

20 According to Keith Sagar, the single most important influence which Ted Hughes offered Sylvia Plath in the early years of their relationship was 'to supply her with a fully worked-out belief in the poetic mythology of Robert Graves's *The White Goddess*' (Sagar, 1978: 11).

21 Was Hughes using Plath's menstruation as inspiration here? Catherine Thompson proposed that Plath suffered from severe cases of premenstrual syndrome. In 'Dawn Poems in Blood: Sylvia Plath and PMS' (1990/1991) she theorised that Plath's mood swings, severe depressions, and ultimately her suicide, were traceable to the poet's menstrual cycles and the hormonal disruptions caused by PMS. See Thompson, '"Dawn Poems in Blood": Sylvia Plath and PMS', *TriQuarterly* 80 (1990/1991): pp. 221–249.

22 In the same way it is interesting to note that many of the *Crow* poems also present different accounts of the same events, such as Crow's creation.

23 *The Iron Man* was first published by Faber and Faber in 1968, and illustrated by Andrew Davidson. It has recently been republished by Walker Books (2010) in collaboration with Faber and Faber and features the drawings of Laura Carlin, which integrate artwork with the text in a far more compelling and resonant way than in the first edition.

24 Up until the third English edition, published in 1985 and the Second American edition, published in 1988, the dedication read as 'To Frieda, Nicholas and Shura'. Curiously, after Hughes's death, the name of Shura appears to have been removed from the dedication page, so that both the 2005 edition with Tom Gauld's illustrations, or the more recent 2010 and 2015 special editions with drawings of Laura Carlin both read as 'To Frieda and Nicholas'.

25 In America the title was changed to *The Iron Giant* so as to avoid confusion with Marvel Comics' character Iron Man. Although the book was not as widely recognised

in America, in 1999 Warner Brothers released an animated film, *The Iron Giant*, based on the novel, directed by Brad Bird and with Peter Townshend credited as executive producer, which became an instant success. Pete Townshend, the rock musician and lead guitarist of 'The Who', also released a musical rock adaptation: *The Iron Man: A Musical*, although critical reaction to the record (and sales) were mostly negative. It opened at the Young Vic, in London, on 25 November, 1993, and was presented as a Christmas show for children.

26 See http://petetownshend.net/musicals/the-iron-man.

27 As previously revealed, one of the rituals that Hughes often re-enacts in his writing is the ritual of death and rebirth, the ultimate sacrifice needed to sustain life. A clear example of this can be seen in 'Orpheus'. Both *The Iron Man* and 'Orpheus' read as examples of Hughes's ability to tell visionary stories about heroic journeys. However, in the case of 'Orpheus' there is no happy ending, whilst *The Iron Man*, by confronting danger and offering himself as a sacrifice to restore life on earth, reads as a heroic figure.

28 That Hughes should choose the unusual name 'Hogarth' for the little boy has been discussed by both Lissa Paul and Ann Skea. According to Paul the enigmatic name could come from the eighteenth-century artist William Hogarth who was famous for his caricatures of social and political institutions and evils of his time (2000: 210). On the other hand, Skea considers that Hughes 'chose the name Hogarth for his hero very carefully, with special attention to the sounds and the vibrations which are made when the syllables of the name are spoken' (2008: 3). Whilst both views are interesting, as Paul has recently pointed out, given the fact that 'Fairy tale heroes typically have generic names' (2016: 20), there is also the possibility that Hughes was perhaps simply looking for a popular local name, like that of Hogarth, typically associated with the North of Yorkshire. In Old English 'hog-harth' was the keeper of cattle and pigs and is also as a dialect word for 'lamb enclosure'. In any case, the name Hogarth appears as one that defines strength.

29 See Kerslake (2016) 'Beyond Rachel Carson's "Sense of Wonder": Advocating Environmental Awareness in Ted Hughes's Writing for Children', for a more detailed analysis which explores the common strand between Carson's writing and that of Hughes in the sphere of his children's writing and argues that Hughes goes beyond Carson's initial 'sense of wonder' to attempt a more ambitious environmental awareness in his readers.

30 The ecofeminist use of the cyborg derives from Donna Haraway who coined the term 'natureculture' in her ground-breaking 'Manifesto for Cyborgs: Science, technology, and socialist feminism in the 1980s', 1985.

31 Both flowers appear out of season in the novel: the foxglove typically flowers in late summer and appears here in April, months too early, whilst the snowdrop belongs to the cold winter, and is months too late.

32 The original letter can be found at the Emory archive: MSS 644, Box 54, FF1, Robert W. Woodruff Library, Emory University, Atlanta, Georgia, USA.

Archive material

Add MS 88918/8/5. Children's Literature Drafts (The British Library, Western Manuscripts, London, 21 June 1979).

Add MS 88918/8/2. Unpublished Creation Tales Material (The British Library, Western Manuscripts, London, c.1980s–1990s).

MSS 644, Box 54, FF1. January–March 1992, Series 1: Correspondence, 1959–1997, Subseries 1.5: Ted Hughes' Correspondence 1957–1997 from Ted Hughes Papers, 1940–1999 (Robert W. Woodruff Library, Emory University, Atlanta, Georgia, USA).

MSS 644, Box 139. Series 3: Sylvia Plath, Subseries 3.2: Prose from Ted Hughes Papers, 1940–1999 (Robert W. Woodruff Library, Emory University, Atlanta, Georgia, USA).

MSS 644, Box 115. Series 2: Writings by Ted Hughes, circa 1949–1990, Subseries 2.4d: Essays, Reviews, Talks from Ted Hughes Papers, 1940–1999 (Robert W. Woodruff Library, Emory University, Atlanta, Georgia, USA).

References

Anderson, Nathalie, 'Ted Hughes and the Challenge of Gender', in Keith Sagar (ed.), *The Challenge of Ted Hughes* (London: Macmillan, 1994), pp. 91–115.

Armitage, Andrew, 'The Birthday Letters Myth' (unpublished thesis, Victoria University of Wellington, 2010).

Bate, Jonathan, *Ted Hughes: The Unauthorised Life* (London: William Collins, 2015).

Bentley, Paul, *The Poetry of Ted Hughes: Language, Illusion & Beyond* (London and New York: Longman, 1998).

Bookchin, Murray, *The Ecology of Freedom: The Emergence and Dissolution of Hierarchy* (Palo Alto, CA: Cheshire Books, 1982).

Brain, Tracy, 'Hughes and Feminism', in Terry Gifford (ed.), *The Cambridge Companion to Ted Hughes* (Cambridge: Cambridge University Press, 2011), pp. 94–106.

Caracciolo, Peter L., 'Buddhist Teaching Stories and Their Influence on Conrad, Wells and Kipling: The Reception of the Jataka and Allied Forms in Victorian Culture', *The Conradian*, 11:1 (8 May 1986), pp. 24–34.

Carson, Rachel, *Silent Spring* (Boston, MA: Houghton Mifflin, 1962; London: Penguin, 1999).

Carson, Rachel, *The Sense of Wonder* (New York: Harper & Row, 1965; New York: HarperCollins, 1998).

Drangsholt, Janne Stigen, 'Hughes and Gender', in Terry Gifford (ed.), *New Casebooks: Ted Hughes* (London: Palgrave Macmillan, 2015), pp. 85–99.

Gifford, Terry (ed.), *Ted Hughes* (London: Routledge, 2009).

Gifford, Terry (ed.), *The Cambridge Companion to Ted Hughes* (Cambridge: Cambridge University Press, 2011).

Golding, William, 'Fable', in Margaret Meek, Aiden Warlow and Griselda Barton (eds), *The Cool Web: The Pattern of Children's Reading* (London: The Bodley Head, 1977), pp. 226–240.

Graves, Robert, *The White Goddess: A Historical Grammar of Poetic Myth* (London and Boston: Faber & Faber, 1948; New York: Farrar, Straus & Giroux, 1966).

Haraway, Donna, 'Manifesto for Cyborgs: Science, Technology, and Socialist Feminism in the 1980s', *Socialist Review*, 15:2 (1985), pp. 65–107.

Hovender, Fiona, Linda Janes, Kath Woodward and Gill Kirkup (eds), *The Gendered Cyborg: A Reader* (London: Routledge, 2000).

Hughes, Ted, *How the Whale Became* (London: Faber & Faber, 1963).

Hughes, Ted, *Nessie the Mannerless Monster* (London: Faber & Faber, 1964).

Hughes, Ted, *The Iron Man* (London: Faber & Faber, 1968).

Hughes, Ted, 'Myth and Education', *Children's Literature in Education*, I (March 1970), pp. 55–70.

Hughes, Ted, *Cave Birds* (London: Faber & Faber, 1978).

Hughes, Ted, 'Baboons and Neanderthals: A Rereading of the Inheritors', in John Carey (ed.), *William Golding: The Man and His Books: A Tribute on His 75th Birthday* (London: Faber & Faber, 1986), pp. 161–168.

Hughes, Ted, *Ffangs the Vampire Bat and the Kiss of Truth* (London: Faber & Faber, 1986).

Hughes, Ted, *Tales of the Early World* (London: Faber & Faber, 1988).

Hughes, Ted, 'The Interpretation of Parables', *Signal: Approaches to Children's Books*, 69 (September 1992), pp. 147–152.

Hughes, Ted, *Shakespeare and the Goddess of Complete Being* (London: Faber & Faber, 1992).

Hughes, Ted, *The Iron Woman* (London: Faber & Faber, 1993).

Hughes, Ted, *Winter Pollen* (London: Faber & Faber, 1994).

Hughes, Ted, 'Where is the Key to the Universe?', in Michael Morpurgo, *Muck and Magic: Stories from the Countryside* (London: Heinemann & Mammoth, 1995), pp. 41–56.

Hughes, Ted, *Birthday Letters* (London: Faber & Faber, 1998)

Hughes, Ted, *Collected Poems* (London: Faber & Faber, 2003).

Hughes, Ted, *The Dreamfighter and Other Creation Tales* (London: Faber & Faber, 2003).

Hughes, Ted and Jim Downer, *Timmy the Tug* (London: Thames & Hudson, 2009).

Jacques, Zoe, *Children, Literature and the Posthuman* (London: Routledge, 2015).

Kazzer, Claas, 'Earth Moon: Ted Hughes's Books for Children (& Adults)', in Joanny Moulin (ed.), *Ted Hughes: Alternative Horizons* (London: Taylor & Francis, 2004), pp. 111–124.

Kazzer, Claas, 'Family Relations – Traces of a Cosmology in Ted Hughes's Creation Tales'. Unpublished paper presented at Ted Hughes Conference at Emory University in Atlanta in 2005. Available online: www.ted-hughes.info/criticism/online-articles/kazzer-claas/family-relations.html.

Kerridge, Richard, 'Ecocritical Readings', in Terry Gifford (ed.), *New Casebooks: Ted Hughes* (London: Palgrave Macmillan, 2015), pp. 176–190.

Kerslake, Lorraine, 'Beyond Rachel Carson's "Sense of Wonder": Advocating Environmental Awareness in Ted Hughes's Writing for Children', *The Ted Hughes Society Journal*, V:I (2016), pp. 65–73.

Lykke, Nina and Rosa Braidotti (eds), *Between Monsters, Goddesses and Cyborgs: Feminist Confrontations with Science, Medicine, and Cyberspace* (London: Zed Books, 1996).

Middlebrook, Diane, *Her Husband* (London: Little, Brown, 2004).

Morrison, Blake, 'Man of Mettle', *Too True* (London: Granta, 1999), pp. 157–172.

Orwell, George, *Animal Farm* (London: Penguin, 1989).

Paul, Lissa, 'The Return of the Iron Man', *The Horn Book Magazine* 76:2 (March/April 2000), pp. 218–225.

Paul, Lissa, 'Ted Hughes and the Environmental Imagination: Brought to You by the Letter R', *The Ted Hughes Society Journal*, V:I (2016), pp. 18–24.

Philip, Neil, 'The Shawl of the Beauty of the World: The Children's Books of Ted Hughes', *Signal: Approaches to Children's Books*, 34:100 (2003), pp. 191–202.

Plath, Sylvia, *Collected Poems*, edited by Ted Hughes (London: Faber & Faber, 1981).

Reddick, Yvonne, *Ted Hughes: Environmentalist and Ecopoet* (Basingstoke: Palgrave Macmillan, 2017).

Reid, Christopher (ed.), *Letters of Ted Hughes* (London: Faber & Faber, 2007).

Roberts, Neil, *Ted Hughes: A Literary Life* (Basingstoke: Palgrave Macmillan, 2006).

Roszak, Theodore, *Where the Wasteland Ends* (Garden City, NY: Doubleday, 1973).

Sagar, Keith (ed.), *The Art of Ted Hughes* (Cambridge: Cambridge University Press, 1978).

Sagar, Keith (ed.), *The Laughter of Foxes* (Liverpool: Liverpool University Press, 2006).
Sagar, Keith (ed.), *Poet and Critic: The Letters of Ted Hughes and Keith Sagar* (London: British Library, 2012).
Skea, Ann, *The Poetic Quest* (Armidale, NSW: University of New England Press, 1994).
Skea, Ann, 'The Iron Man', 2008. Web. 11 October 2013. http://ann.skea.com/NQ6iron.htm.
Thompson, Catherine, '"Dawn Poems in Blood": Sylvia Plath and PMS', *TriQuarterly*, 80 (Winter 1990/1991), pp. 221–249.
Ushua, V. T., *Modern English Literature: The Real and the Imagined: The Poetic World of Ted Hughes* (Jaipur: Mangal Deep Publications, 1998).
Vanderbilt, Gloria, 'Review of *How the Whale Became* by Ted Hughes', *New York Times Book Review* (8 November 1964), p. 28.
White, Lynn, Jr., 'The Historical Roots of Our Ecologic Crisis', *Science*, 155 (10 March 1967), pp. 1203–1207.
Wordsworth, William, *The Oxford Authors: William Wordsworth*, edited by Stephen Gill (Oxford: Oxford University Press, 1984).

Conclusion

Hughes's poetic journey is both a celebration of life and, more importantly, it conveys the recognition of our immersive, inescapable relationship with nature, immortalised in the lines of his poem 'That Morning', inscribed on the poet's memorial stone at the foot of T. S. Eliot's grave at Westminster Abbey:

> So we found the end of our journey,
> So we stood alive in the river of light,
> Among the creatures of light, creatures of light.
> (*River*, 1983; *CP*: 664)

A vision which expresses, in Sagar's words, 'the divine harmony of matter and spirit ... as if this were no longer a fallen world' (2006: 167). Behind Hughes's poetry pulsates that same idea of wholeness, one that is achieved at the end of Hughes's shamanic journey – a journey where the poet finally finds light emerging from the darkness.

This book has attempted to explore not only the importance of Hughes as a children's writer but more significantly the imaginative power of Ted Hughes's children's writing and the healing function of nature that emerges so clearly in his work. Following this, *The Voice of Nature* argues that Hughes writes for a world that has lost its balance. It explores the way in which the poet articulates our vital relationship with nature, speaking not only as a voice for nature, but, as it were, through nature itself in the realm of his children's fiction. Hughes's children's writing engages with our ecological crisis by exploring the relationship between humans and nature and showing how our present ecological crisis is the fruit of an anthropocentric dominating view of modern society that has alienated us from nature and the inner world of the human psyche. As Sagar argues in *Literature and the Crime Against Nature* (2005), all literature has engaged with the issue of the relationship between our species and nature:

> Ecology needs imagination, and imagination's most articulate expression, literature ... The literary imagination connects all the severed halves – inner and outer, self and other, male and female, life and death, man and Nature. Every metaphor is a stitch in the suture.
>
> (371)

In the case of Hughes, this holistic sense of literature represents the poet's own mythical storytelling and is key to understanding his healing quest. It is by connecting, as it were, 'the severed halves' that Hughes seeks to re-establish the balance and harmony between humans and nature. However, despite underlying his entire poetic scheme, this aspect has remained unrealisable and problematic in his adult work, since redemption is never fully obtained, even in 'That Morning'.

Hughes's critique of humanity suggests that ecological problems also stem from social problems. As Hughes himself states: 'We have no problem in becoming aware that we are destroying our planet and our life on it. Our problem is the very great difficulty of becoming aware in such a way that we do something about it' (1993: vii). However, as this book has shown, it is in the realm of his children's writing where he successfully explores how human alienation from nature can be understood and resolved by offering a solution to the problem.

Hughes believed that not only could nature be saved but that humanity could change if it was willing to rethink its attitudes towards its place in the web of life. Through the figure of the 'Goddess of Complete Being' and the power of myth he was able to explore the primal energies of the natural world and the creative-destructive forces of the universe in his children's literature more fully than in that for adults. His writing for children embodies a hidden autobiographical self, one which is closely related to his cathartic quest of healing throughout his imaginative writing. The poet's empathy towards the child's imagination is also closely linked to his personal healing quest and his ongoing project to write in a 'lingua franca' for a child audience, but for one in which adults could 'overhear'. Hughes's poetic journey is an ongoing venture to reconnect the reader (both the adult and the child) to what Rachel Carson called 'a sense of wonder' (see Carson, 1965). This reveals that by writing 'within hearing' (Cushman, 1983: 255) of adults and not just children, as Cushman asserts, Hughes may have also been suggesting that the adult listener could be healed by, and as, the child reader.

In the Introduction the question was raised of what an ecocritical reading of Hughes's children's writing could reveal. By extending ecocriticism into the realm of children's literature, it can provide a clear insight into the problematic engagement between the human and non-human and help to show how humans and nature should (and indeed, must) live in partnership. Hughes's children's writing can be employed as a particularly useful tool to offer insights into underlying ecological issues and educating environmental awareness in young children. Not only is Hughes's children's writing sensitive to environmental issues but there is a much clearer sense of ecological healing portrayed throughout his writing for children. This sense of ecological healing is largely a holistic one. The references from deep ecology, ecofeminism and ecopsychology that *The Voice of Nature* draws on attempt to clarify the way Hughes conceives the relationship between human/non-human, highlighting the need to address the questioning of what it means to be human alongside that of other beings in Abram's 'more-than-human world' (1996).

Hughes's children's poetry reads as an instrument to raise ecological awareness and connect with the reader asking them to understand the relation that lies between our familiar local nature and the web of nature to which we belong. As an eco-poet his poetry responds to the modern divorce between humanity and the rest of nature, exposing Western culture's Christian error in teaching how human beings have disturbed nature and the lives of animals, whilst showing the reader how to look at nature from the point of view of the 'other', educating a sense of respect towards the environment. That his poetry works by articulating a sensory affinity between ourselves and nature in order to recover what Carson called 'a sense of wonder' may perhaps be taken as an ecological trait in itself. However, it is by instilling an appreciation of the natural world through his animal poetry, and reconnecting us with the natural world through his ongoing shamanic quest of healing that Hughes's poetic voice works best. This can be seen in the fact that by the end of the 1970s there is a clear change of mood in much of his poetry, given that Hughes was convinced that poetry had a positive, healing effect on readers, even if it meant dealing with themes such as death and suffering. Not only do his children's poems realistically provide his younger audience with information about the natural world and also help to educate and teach respect towards it, but they also raise questions (if only indirectly at times) about important ethical issues such as farming, industrialisation, pollution, hunting or keeping wild animals.

Hughes's healing quest is of course largely informed by his understanding and awareness of mythology together with the importance he places on storytelling. Myth, folk and fairy tales serve Hughes as a symbolic way of presenting the different means by which human beings cope with the world in which they live and allow Hughes to connect his characters and locations to universal truths and themes. Hughes regarded stories as the blueprint for our imagination. As he declared in 'Myth and Education', they are 'little factories of understanding. New revelations of meaning open out of their images and patterns continuously' (*WP*: 41). As both a myth-maker and a storyteller, Hughes created stories that captured the essence of experience through the affective function of language. Following this, Hughes's version of Orpheus reads as his most redemptive play and simultaneously enacts the cathartic sense of healing for Hughes's persona following the loss of Sylvia Plath, whilst both *The Iron Man* (1968) and *The Iron Woman* (1993) are his most successful healing myths. *The Iron Man* reads as a personal healing myth for Hughes and his children, whilst *The Iron Woman* enacts Carson's sense of wonder and is one of Hughes's most powerful educational interventions on behalf of water quality and public health as a vehicle for a wider environmental awareness. By advocating environmental consciousness and education together with respect for the other, throughout his children's writing and work, Hughes hoped to educate future generations. Both physically and spiritually, he has brought us closer to healing the gap between our inner and outer nature – or what he called 'Mother Earth' – than any other poet.

Of course, like all mortals, Hughes was not a perfect human being, and his life will always remain deeply controversial, largely due to the tragedies in his personal life. Despite the many contradictions, encompassing the beauty and the horror that abound in his life and work, this study has focused on the artistic and environmental achievements of Hughes's children's literature. In relation to the apparent conflicts and contradictions that emerge in Hughes's persona, there is an important underlying aspect that remains to be explored further: his attitude towards hunting and fishing and the ethical questions that lie behind it. Although in his writing for children Hughes shows empathy towards the animals hunted, future studies should attempt to investigate this connection between the hunter and the hunted and explore Hughes's somewhat contradictory position on the subject of hunting as well as the question of the ethics of fishing in relation to writing for children.

There are other areas of fundamental importance to Hughes's career that have remained largely neglected in relation to his children's writing. One of the most interesting areas of development for future studies lies in exploring the depth of Hughes's educational achievements. Given the scope of this study it has only been possible to briefly discuss projects such as 'Farms for City Children', his wish to set up a tree planting scheme with schools in Devon or his concern for the Atlantic Salmon Trust (AST) in the mid-1980s,[1] and The Sacred Earth Drama Trust in the 1990s, which aimed to make children spokespeople for the environment. During the early period of Hughes's writing for children and throughout his life, his own convictions together with his ecological concerns not only underlined the thinking behind his poetic voice but also led to his life-long commitment and support to a number of educational projects related to encouraging children to write poetry themselves. One of Hughes's less acknowledged but serious commitments was his work as a judge in the *National Children's Poetry Competition* of the *Daily Mirror*, which later became the *W. H. Smith National Literary Competition*.[2] The fact that he dedicated over 30 years of his life to such a project suggests the importance that Hughes felt children's writing competitions held and the potential of young writers. Indeed, Hughes's life-long commitment and his impact on different organisations and projects was a key factor in their success, as in the case of the Arvon Foundation, a creative writing residential workshop set up as a reaction against the dogmatic approach of teaching poetry in schools and established in 1968 by John Moat and John Fairfax in Devon.[3] These are areas that would benefit from further examination, in line with the importance that Hughes gave to engaging with children throughout his working life in order to develop their creativity and imagination and his particular insight into their sensibility.

On 28 October 1998, Hughes's own physical and poetic journey came to an end, after losing his fight against colon cancer. Speaking at his funeral, Hughes's close friend, the poet Seamus Heaney said that:

> He was a tower of tenderness and strength, a great arch under which the least of poetry's children could enter and feel secure. His creative powers

> were, as Shakespeare said, still crescent. By his death the veil of poetry is rent, and the walls of learning broken.
>
> (In Boyanowsky, 2009: 195)

His poetic journey towards reconciliation is best summed in his own words when he wrote that 'Every new child is nature's chance to correct culture's error' (*WP*: 149) underlying his belief in imagination as the most powerful tool for healing. What remains clear today is that his work continues to bring us light into a much darkened world, and judging from the immense variety and scope of the poet's work, it will continue to do so and be worthy of study in the years to come.

Notes

1 Hughes was a friend and supporter of the AST, to which he dedicated the poem, 'The Best Worker in Europe', in 1985.
2 Hughes became a judge on the Advisory Panel in the mid-1960s, assuming chairmanship in 1978, until the competition was abandoned by W. H. Smith in 1997.
3 In 1975, when Hughes learned that Arvon was looking for a northern base as a centre he offered them the use of his home Lumb Bank, and helped financially with the renovation, partly from the sale of his manuscripts. Later in 1989 Arvon bought Lumb Bank from Ted and Carol Hughes, with help from the Arts Council. Not only did Hughes offer financial support, but in 1982, together with Seamus Heaney, Philip Larkin and Charles Causley, he worked as a judge for the *Observer*/Arvon foundation Poetry Competition, to help raise money and save the foundation from collapsing financially.

References

Abram, David, *The Spell of the Sensuous: Perception and Language in a More-than-Human World* (New York: Vintage Books, 1996).
Boyanowsky, Ehor, *Savage Gods, Silver Ghosts: In the Wild with Ted Hughes* (Vancouver: Douglas & McIntyre, 2009).
Carson, Rachel, *The Sense of Wonder* (New York: Harper & Row, 1965; New York: HarperCollins, 1998).
Cushman, Keith, 'Hughes' Poetry for Children', in Keith Sagar (ed.), *The Achievement of Ted Hughes* (Manchester: Manchester University Press, 1983), pp. 239–256.
Hughes, Ted, *The Iron Man* (London: Faber & Faber, 1968).
Hughes, Ted, *River* (London: Faber & Faber, 1983).
Hughes, Ted, *The Iron Woman* (London: Faber & Faber, 1993).
Hughes, Ted, *Winter Pollen* (London: Faber & Faber, 1994).
Hughes, Ted, *Collected Poems* (London: Faber & Faber, 2003).
Sacred Earth Drama Trust, *Sacred Earth Dramas*, Foreword by Ted Hughes (London: Faber & Faber, 1993).
Sagar, Keith, *Literature and the Crime Against Nature* (London: Chaucer Press, 2005).
Sagar, Keith (ed.), *The Laughter of Foxes* (Liverpool: Liverpool University Press, 2006).

Index

For Product Safety Concerns and Information please contact our EU representative GPSR@taylorandfrancis.com Taylor & Francis Verlag GmbH, Kaufingerstraße 24, 80331 München, Germany